Environmental Restoration: Hope from the Frontiers of Science

Proceedings of the Second International Conference on Science and God

Global Online Conference
April 22 - 23, 2021

Hyo Jeong International Foundation
For the Unity of the Sciences Press

This Proceedings volume contains content delivered at the
Second International Conference on Science and God (ICSG II).
This book is the Second Edition.

Book design and cover photo by Gil Roschuni

Library of Congress Control Number: 2023909415
ISBN: 978-1-947499-11-9

Published by

Hyo Jeong International Foundation for the Unity of the Sciences Press

3600 New York Ave. NE
Washington DC 20002
USA

324-211 Misari-ro, Seorak-myeon
Gapyeong-gun, Gyeonggi-do
Republic of Korea

Table of Contents

Preface

This book contains the proceedings of the Second International Conference on Science and God (ICSG II), held on April 22–23, 2021 (Korea time), using a virtual platform covering the globe. The conference, sponsored by the Hyo Jeong International Foundation for the Unity of the Sciences (HJIFUS), was a joint project of HJIFUS offices in the United States and Korea, with support from related offices in Japan and Europe.

HJIFUS was founded on the vision and teachings of Rev. Dr. Sun Myung Moon and Rev. Dr. Hak Ja Han Moon. Its mission is to identify, promote, and implement the most effective solutions to environmental problems on a worldwide scale. As an important step in this process, HJIFUS has been sponsoring the International Conference on the Unity of the Sciences (ICUS) annually since 2017. The main focus of the ICUS series has been on examining technologies derived from conventional scientific approaches to help resolve environmental issues. In 2020, HJIFUS launched the ICSG series to broaden our search by exploring new approaches based on frontier science.

ICSG II was unusual in that it was convened online. At that time, COVID-19 had become a serious pandemic, claiming numerous lives around the world. Consequently, international travel and social gatherings were highly restricted, and in-person meetings had to be ruled out. Fortunately, electronic communications technology was sufficiently developed to make it possible for individual participants to form a network of connections through their Internet-enabled devices, such as laptop computers and mobile phones. To run the virtual platform across multiple time zones around the world, we engaged the services of iPeaceTV in Korea, which has state-of-the-art equipment and a highly skilled, dedicated staff.

The overall theme for ICSG II was, "Environmental Restoration: Hope from the Frontiers of Science." The organizing committee consisted of scientists and scholars with prior experience in the organization of international conferences on environment-related issues, and those who could help identify appropriate themes, topics, and qualified speakers. There were 70 registered attendees representing 12 countries, including trained researchers who were willing to expand the scope of their investigations outside the mainstream box. Professor Luc Montagnier, an eminent medical researcher and recipient of the 2008 Nobel Prize in Physiology or Medicine, served as co-chair for the conference.

ICSG II had the following structure: an opening session, including a keynote address, followed by three academic sessions. Each academic session consisted of presentations on two topics, with two commentaries on each topic, and a discussion in plenary format. The conference ended with a closing session.

Building on the foundation laid by ICSG I, this conference challenged the materialist philosophy underlying much of Western scientific thought and paved the way for greater awareness and acceptance of the incoming postmaterialist era. Starting with the keynote address, the speakers provided evidence showing that human consciousness transcends the tangible, material world that we can access using our physical senses, and that the fundamental nature of reality is nonmaterial. In addition, it was shown that, on a deep level, we are interconnected with one another and with our natural environment. Another presentation tackled the question of how human consciousness may interact with the physical brain. Taken together, these presentations awaken us to recognize the reality of higher, spiritual realms, as well as the significance of protecting and restoring the environment of which we are a part.

The Appendix of this book provides information about the conference organizing committee and list of attendees, as well as some background on the founders of ICSG and HJIFUS. It further includes a short introduction to HJIFUS, with the boards of directors and contact information for our worldwide offices.

I wish to thank all those who helped make ICSG II a remarkable success, including the conference co-chair, session chairs, presenters, other participants, organizing committee members, and staff. I also thank all who invested their efforts to produce this volume although the pandemic delayed its publication. ◇

Dr. Douglas D.M. Joo
Chairman, ICSG II and HJIFUS

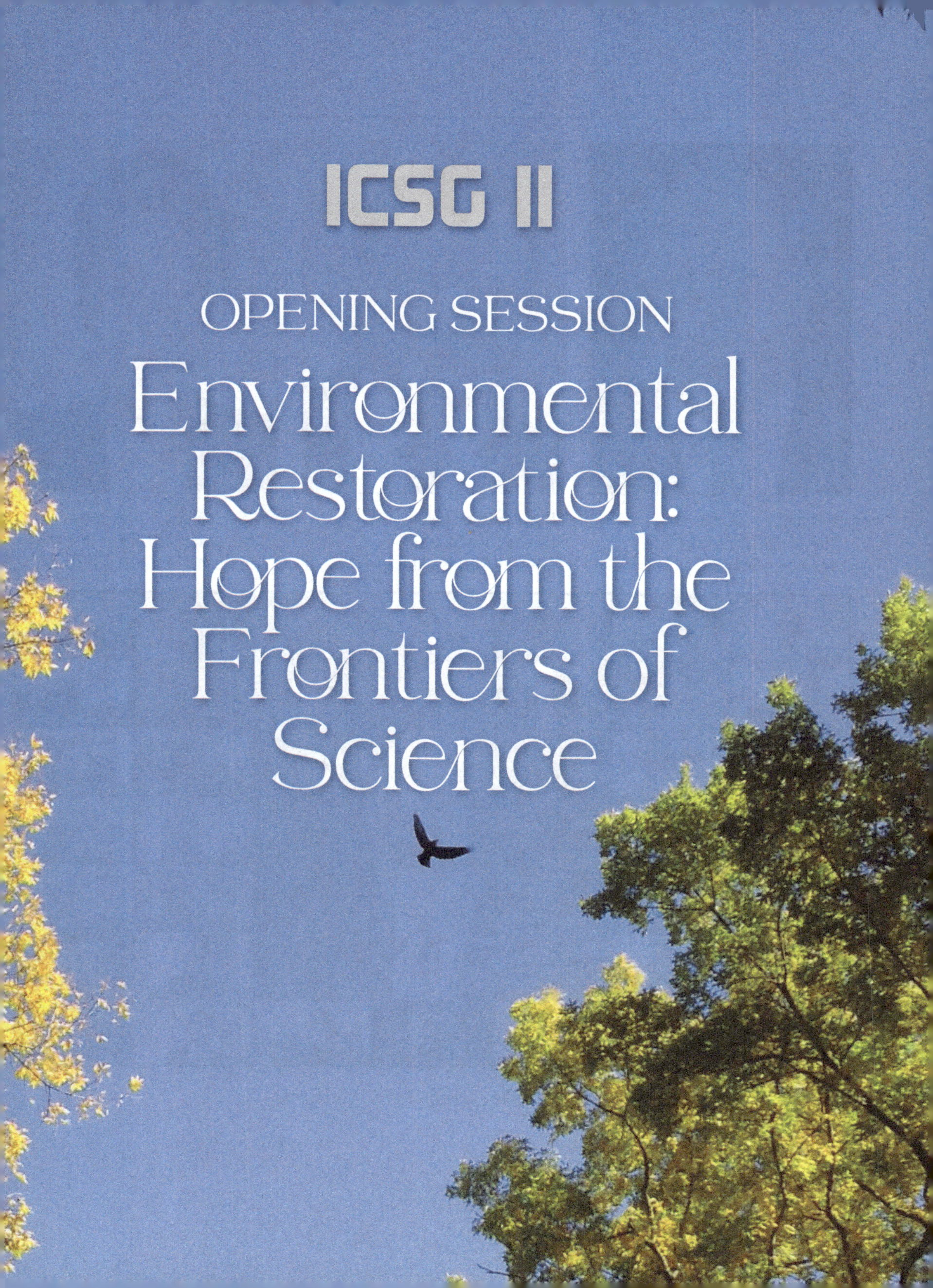

ICSG II

OPENING SESSION

Environmental Restoration: Hope from the Frontiers of Science

Dr. Douglas D.M. Joo, Chair of ICSG II, delivers the Welcoming Remarks.

Prof. Luc Montagnier, Co-Chair of ICSG II, gives the Opening Remarks.

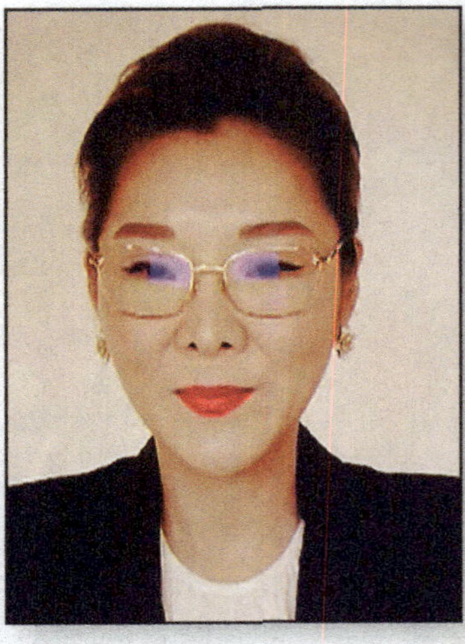

Dr. Sun Jin Moon gives the Founder's Address on behalf of Rev. Dr. Hak Ja Han Moon, Cofounder of ICSG.

Conference participants listen to the Keynote Address given by Prof. Gary Schwartz.

Opening Session

Welcoming Remarks

Dr. Douglas D.M. Joo
Chair, ICSG II

Respected Co-Chair Professor Luc Montagnier, Scientists, Scholars, Ladies and Gentlemen:

On behalf of the Hyo Jeong International Foundation for the Unity of the Sciences (HJIFUS), I am very pleased to welcome you to the Second International Conference on Science and God (ICSG II). The theme of this conference is, "Environmental Restoration: Hope from the Frontiers of Science."

As you may know, over the past four years, HJIFUS has been sponsoring a conference series called the International Conference on the Unity of the Sciences (ICUS), focusing on conventional scientific approaches toward finding solutions to environmental problems. Last year, we held the first ICSG meeting in Seoul, Korea, to open the door to unconventional, "frontier science" approaches, as a way to broaden our search for the most promising solutions. Judging from the reflections of the participants, it was a very successful meeting. It showed us that we have settled into a new era in the history of science.

Since then, however, circumstances have changed dramatically. A previously unknown coronavirus spread quickly around the globe, resulting in a serious pandemic that was named COVID-19. Millions of people have lost their lives, and many more have been suffering either directly or indirectly from the effects of the disease.

At the same time, the pandemic has underscored the fact that we are all interconnected with one another and with our natural environment. It has demonstrated the need to reaffirm our commitment to the principles of interdependence, mutual prosperity, and universally shared values. We need to value the world of nature and work together to restore our environment for the benefit of everyone on our planet.

Thus, despite the pandemic, we recognized the importance of convening ICSG II this year, to build on the success of last year's meeting. Fortunately, we could turn to modern technology, which provides us with the means to hold the conference using a virtual platform, transcending the limits of space and time. I respect and congratulate all of you, eminent scientists and scholars, for leading the way in overcoming the threat of the pandemic, challenging the difficulties imposed by geographical separation and differing time zones, while simultaneously engaging in pioneering research that goes beyond the boundaries of established, conventional science.

Furthermore, I wish to express my gratitude to the iPeaceTV technical team and time zone coordinators for the tremendous job they have been doing in setting up and operating the virtual platform, to enable the smooth flow of this historic event.

ICSG II has been designed to cover a range of stimulating topics. It will start by looking at the flow of science through different eras, going from tangible to intangible dimensions, and note the radical changes in scientific concepts about the nature of the universe. Another session will examine how amazing ideas and technologies arising from the cutting edge of science offer startling new ways to resolve environmental problems and restore health. The final session will assess how human consciousness interacts with the brain, and whether investigations at the frontiers of science can lead to a deeper understanding of the First Cause of the universe.

As you may know, April 22 is observed as Earth Day, and many regard April as Earth Month. I am happy to announce that this month, HJIFUS is launching an online, bimonthly publication, *The Earth & I.* Its purpose is to educate the general public on a wide range of environmental issues, and to help unite the people of the world in appreciating and caring for the environment. In addition to conveying the latest news on the environment, it will cover a variety of issues, ranging from ecosystems and energy to food, water quality, and climate change. It will combine articles on serious environmental challenges with enjoyable articles about the world of nature.

A few days from today, on April 27, is the 61st holy wedding anniversary of our founders, Rev. Dr. Sun Myung Moon and Rev. Dr. Hak Ja Han Moon. So, this is also a time to appreciate them for sharing their love for nature and to celebrate their lifelong investment to improve the quality of the environment and the quality of life for all humankind.

God has given us many blessings, in terms of both human and natural resources, as well as technological advances. I hope we can use these resources well and share them with others, to break down all barriers that are blocking us around the world.

I wish you a very inspiring and productive meeting.

Thank you very much. ◊

About the Speaker

- Chair, Hyo Jeong International Foundation for the Unity of the Sciences
- Former Chair and President of *The Washington Times* and United Press International, and several other international news publications under News World Communications, as well as AmericanLife TV Network and other electronic media
- Former President, World Media Association, The Washington Times Foundation, American Family Coalition, and Times Aviation International
- Former International President of the Youth Federation for World Peace
- Former Co-Chair of the Board of Trustees of the University of Bridgeport, Connecticut
- An honorary doctorate in Political Science, Sun Moon University; M.Ph. in Political Science, George Washington University

Opening Session

Opening Remarks

Professor Luc Montagnier
Co-Chair, ICSG II

Madame Moon, members of the Moon family, Dr. Joo, and our friends in the Unification Church, thank you for arranging this Second International Conference on Science and God. To our friends and colleagues from around the world, I thank you for joining us.

More than ever, God and science are both needed to overcome the problems of these complex times. God and science are part of our lives, for all of us at this conference. Hopefully, the world will soon realize that God and science are necessary for everyone to deal with the challenges we currently face—challenges that may be even more serious in the future.

I would like to take a moment to remember with respect and friendship our colleague, Dr. Mario Molina. Speaking on behalf of participants of the International Conference on the Unity of the Sciences, I wish to say that we were privileged to have had him with us in Seoul over these last several years. The world as a whole was privileged to have had the brilliance of Dr. Molina bring to our attention serious threats to our environment. We shall never forget Dr. Molina. We shall be eternally grateful for his selfless dedication to improving environmental conditions for all of humanity.

One of my fields of expertise over the last 60 years has been RNA viruses. Therefore, the COVID-19 pandemic is not a total mystery to me. It is a complex problem afflicting our civilization. It presents many open questions, the answers to some of which may come from God.

Adding to the difficulties and confusion of this last year and a half is the spread of "fake news." The average citizen, a non-scientist, cannot discern the truth in science news and medical news—particularly, pandemic news. During this pandemic, much of what has been presented to the public as medically necessary is actually politically motivated, in the guise of science. The average citizen is disadvantaged by these ruses. The sometimes-corrupt cabal of governments, industry, media, and academic institutions makes it impossible for people to know who and what to believe.

God and science are the basis of life. May we all work to bring God and true science to the forefront.

I join you in looking forward to the upcoming presentations.

Thank you for your attendance and your kind attention. ◇

About the Speaker

- Founder and Director, Fondation Luc Montagnier, Geneva, Switzerland
- Professor Emeritus, Institut Pasteur
- Recipient of the 2008 Nobel Prize in Physiology or Medicine for his contribution to the discovery of HIV, the virus that causes AIDS
- In recent years, has extended his research objectives to the infectious origins of several chronic diseases, including cancer, Alzheimer's and Parkinson's diseases, autism, and Lyme disease
- Concerned about how environmental changes could be involved in these diseases
- MD in Medicine

Opening Session

Founder's Address

Dr. Sun Jin Moon
Representing the Founders

Distinguished Scientists, Scholars, Ladies and Gentlemen:

I feel greatly privileged to greet you at the Opening Session of the Second International Conference on Science and God, or ICSG II. I am especially happy to speak to you on behalf of my parents, the Reverend Dr. Sun Myung Moon and Reverend Dr. Hak Ja Han Moon, whose vision and work form the basis of this conference series.

My mother described the purpose of this conference as follows:

> *To open the door to innovative applications not yet explainable by conventional science but with the potential to greatly improve the environment. At the same time, this conference offers a look into the history of scientific development.... ICSG is showing that scientific trends are pointing toward a new era of science brought on by new paradigm science and are leading to the First Cause of the universe, thereby tearing down the wall between science and religion.*

ICSG is an extension of the work my parents carried out for decades through founding and supporting the series known as the International Conference on the Unity of the Sciences. In the 2019 meeting of ICUS, my mother stated:

> *Year by year, the deterioration of the global environment is expanding. Climate change, extreme weather conditions, deteriorating air quality, and polluted oceans are now topics of everyday conversations. There is*

growing concern that a great catastrophe is upon us. It is no exaggeration to say that the preservation of our environment is the most pressing and weightiest issue humanity faces today.

When we look back at this past year, however, we cannot help but be shocked by the incredibly challenging, life-changing year we have all been through! As the COVID-19 pandemic spread rapidly around the world, each of us was affected dramatically by lockdowns and curfews, face masks, social distancing, and sanitization measures. Many of us lost people we knew and loved. Every person, whether friend or stranger, was a potential carrier, and we ourselves needed to take special care to prevent spreading the disease to others. Businesses closed; churches, temples, and mosques could not meet. Extended families could not gather for holidays, and many grandparents could not even hug their grandchildren. Worse still, too often, our precious elders died without their families present at their bedsides. Schools closed down, travel was severely restricted, and unemployment skyrocketed. National economies were badly shaken. Hospitals and medical facilities were overwhelmed, and the death toll climbed into the millions around the globe. At last, the development of vaccines against the virus has brought a glimmer of hope at the end of the tunnel.

Fortunately, our technological revolution has also brought us to the stage where we can transcend the restrictions imposed by the pandemic and convene this international conference, spanning different time zones throughout the world, using a virtual platform. Thus, it is a real pleasure to speak to all of you who are participating in this historic event.

As you know, this meeting is sponsored by the Hyo Jeong International Foundation for the Unity of the Sciences (HJIFUS). It was founded in 2017, under the inspiration of my dear mother. Recognizing that environmental challenges were multiplying rapidly on a global scale, she saw the urgent need to resolve them. Of course, she was very aware that there are already numerous organizations that are conducting environmental restoration activities. However, HJIFUS can play a unique role by combining external, technological solutions with the restoration of internal relationships of heart—that is, restoration of the originally intended relationships between humans and nature, and eventually between humans, the environment, and the First Cause, or God.

As far back as 1957, my father spoke about these relationships in the following terms (*Cheon Seong Gyeong*, Book 13, p. 1398):

> *God created all things in the universe, our natural environment, which are absolutely necessary for human prosperity. People are meant to harmonize with nature; we, as the subject partners of love, and nature as the object partner, thriving under our management and perfected in beauty by our creativity. Let us not destroy nature and pollute the environment.... Return to nature and enjoy a life of liberation and complete inner freedom. To love nature is to love God and humanity. When human life resonates with nature, human character can blossom in perfection. The flowers of a true culture of heart, a true artistic world, will bloom.*

Changing our attitude toward nature implies taking responsibility for our environment. As my father stated in 1985 (*Cheon Seong Gyeong*, Book 4, p. 393):

> *Human beings should fulfill our portion of responsibility during the time we are growing to maturity... In this way, we fulfill the condition of having cooperated in God's work of creation and can share the authority of the Creator.*

As we know, the United Nations Environment Programme and other environmental initiatives are doing important work on many of these issues, but they often lack political backing. On the other hand, organizations that are supported by particular governments tend to promote the narrow interests of their home countries rather than putting the interests of the world first. In addition, in my opinion, today's international environmental organizations also lack the spiritual vision that is necessary to inspire people to transcend political and ideological disparities.

Ladies and Gentlemen, we need an international environmental Congress or Council, consisting of experts from around the world, to overcome these obstacles.

HJIFUS is considering how to create it. It is my hope that your deliberations at this conference will contribute significantly to workable plans to do exactly that.

The current activities of HJIFUS include environmental research, hosting environmental conferences, publishing the periodical *The Earth and I,* and developing drinking water supplies for disadvantaged communities in Africa, Southeast Asia, and Latin America. Planned activities include proposing environmental policies for national governments and international organizations, developing new products that benefit the environment and producing and distributing them, organizing social campaigns to promote environmental awareness and train young people, and opening the way for new paradigm science to develop methods to regenerate the environment. With your help, we hope to expand these activities and contribute significantly toward the realization of the vision our Founders have articulated.

This particular conference is focused on innovative approaches and new paradigms that involve spirituality as well as pure science. As my mother has stated (*Mother of Peace,* p. 183):

> *Once in a while, you hear religious people devalue science as having nothing to do with God, and secular people devalue religion as having no practical use. Both sides separate God and science. Both are in error. God wants us to develop science and technology as tools with which we can exercise dominion of love over all things. And that is God's great blessing. We must love nature with the same heart as God and cultivate it for the benefit of humanity. This is God-centered science and technology.*

In conclusion, I want to thank you for your participation. I look forward to learning of your innovative approaches and to many years of fruitful cooperation toward our important mutual goals. ◊

About the Speaker

- Fifth daughter of Rev. Dr. Sun Myung Moon and Rev. Dr. Hak Ja Han Moon
- Senior Vice President, Women's Federation for World Peace International

(continued on next page)

- Chair, Pacific Rim Education Foundation in Kailua-Kona, Hawaii (since 2012)
- Has served the Family Federation for World Peace and Unification, Universal Peace Federation, and Women's Federation for World Peace (since 2000)
- Baccalaureate in psychology from Harvard University
- Inducted into the Alpha Sigma Lambda National Honor Society
- Candidate for Master of Arts degree in psychology from Columbia University
- Received an honorary doctorate in Social Science from Sun Moon University

Opening Session | Keynote Address

Insights into Material and Nonmaterial Reality

Professor Gary E. Schwartz
Psychology, Medicine, Neurology, Psychiatry, and Surgery
The University of Arizona, Tucson, USA

"All matter originates and exists only by virtue of a force... We must assume behind this force the existence of a conscious and intelligent Mind. This Mind is the matrix of all matter.

—Max Planck

Clearly, the universe had meaning and direction ...an unseen dimension behind the visible creation that gives it an intelligent design and that gives life purpose.

—Edgar Mitchell

The title of this keynote address, "Insights into Material and Nonmaterial Reality," was requested by the organizers of the Second International Conference on Science and God. Given the scope and significance of this visionary conference, I have felt a special responsibility to honor the organizers' intent to address the big picture of this unfolding revolution in science and society, as illustrated, for example, by the emergence of postmaterialist theory, research, and applications (for example, see www.AAPSglobal.com).

For over a century, many philosophers, scientists, and other scholars have proposed that we re-envision[1] the existence of a greater reality whose fundamental existence is nonmaterial, namely, the very stuff of Mind itself. They have argued not only that the primacy of Mind and Consciousness reflects a truer description of Reality but also that universal appreciation and acceptance of this understanding are essential for humanity's survival, restoration, and evolution (for example, Schwartz et al. 2020).

As a Western-educated scientist trained in materialist science,[2] I never questioned my materialism-based education prior to my early forties. My unquestioned beliefs about the material foundation of reality were fortified by my nonreligious upbringing and my assumptions that all spiritual theories and experiences reflected some combination of ignorance, superstition, faulty reasoning, and self-confirming biases.

However, the second half of my academic career has radically transformed my understanding and appreciation of nature, the Cosmos, and Mind itself. Although painful at times, what I have learned over the years is that instead of science taking us away from the idea that consciousness is fundamental in the universe, science is inexorably taking us to the primacy of consciousness hypothesis (for example, Schwartz et al. 2020). I have learned that (1) logic and reasoning, and (2) observations and evidence, combined with (3) direct personal experiences triangulate and converge on the thesis that "Mind is the matrix of all matter," as Max Planck proposed.

So, what potential insights might I bring to bear on the nature of material and nonmaterial reality that may be useful in broadening and deepening your appreciation of these matters (no pun intended)? The following discussion combines information that I have published previously, with information I have rarely or never previously shared, either in presentations or in print. I should acknowledge that, to the best of my knowledge, I have never used, and certainly never featured, the words "nonmaterial reality" in a presentation or paper. Hence, to minimize miscommunication, I will briefly explain how I am using this phrase.

The following discussion is organized in four parts:

Part I presents some exemplary insights gleaned from the physics of light that speak to the fundamental existence and nature of a nonmaterial reality.

Part II provides some complementary challenging insights from contemporary research on life after death that illuminate the novel (and often surprising) nature of a greater, nonmaterial reality.

Part III discusses evidence documented in daily life—derived from extraordinarily complex and astronomically improbable serial synchronicities—that offer compelling justification for Max Planck's conclusion that "Mind is the matrix of all matter" and Edgar Mitchell's conclusion that there exists "an unseen dimension behind the visible creation that gives it [the material universe] an intelligent design and that gives life purpose." (For the full quotation, see the sidebar.)

Part IV considers how the emergence of postmaterialist knowledge heralds an opportunity for humanity to make a "quantum jump" in its "Meta-Consciousness," that is, in its "consciousness of Consciousness," extending from our experiences of the material world to nonmaterial reality and to the Source/Divinity itself. Such Meta-Consciousness promises to inspire personal, environmental, and global restoration and evolution.

To facilitate the discussion in being accessible to nonscientists as well as scientists, I have minimized methodological or statistical discussions in this keynote address.

Part I: Insights into the Inherent Nonmaterial Reality of Photons, Energy, and Information

I have had a longstanding interest in the quantum and relativistic properties of light, and I have employed state-of-the-art optical sciences in some of my academic research over the past decade (for example, Schwartz 2010, 2011a, 2021a). As a psychologist, I have been fascinated with how sophisticated quantum physicists (and, by extension, the rest of us) struggle to understand the empirical nature of light and fields of energy and information.

For example, optical scientists observe that, in certain contexts, light behaves as if it is a particle localized in space and, in other contexts, light behaves as if it is a wave distributed in space. Scientists observe that in certain states, light appears to have no mass, and in other states (particularly at the speed of light) it behaves as if it has a tiny amount of mass. Scientists

From Outer Space to Inner Space: An Astronaut's Odyssey

Excerpt from an article by Edgar Mitchell, Apollo 14 astronaut

But there was another aspect to my experience during Apollo 14, and it contradicted the "pragmatic engineer" attitude. It began with the breathtaking experience of seeing planet Earth floating in the vastness of space.

The first thing that came to mind as I looked at Earth was its incredible beauty. Even the spectacular photographs do not do it justice. It was a majestic sight—a splendid blue and white jewel suspended against a velvet black sky. How peacefully, how harmoniously, how marvelously it seemed to fit into the evolutionary pattern by which the universe is maintained.

In a peak experience, the presence of the divinity became almost palpable, and I knew that life in the universe was not just an accident based on random processes. This knowledge came to me directly—noetically. It was not a matter of discursive reasoning or logical abstraction. It was an experiential cognition. It was knowledge gained through private subjective awareness, but it was—and still is—every bit as real as the objective data upon which, say, the navigational program or the communications system were based. Clearly, the universe had meaning and direction. It was not perceptible by the sensory organs, but it was there nevertheless—an unseen dimension behind the visible creation that gives it an intelligent design and that gives life purpose. ◊

observe that light behaves as if it has energy as well as complex patterns of information (that is, frequencies) over space and time.

In physics, energy is defined as the capacity to do work and overcome resistance, and information is described as abstract (mathematical) patterns in space and time. What is important (and often ignored) is that neither energy nor information is described as has having "mass" or being "material." In other words, in a deep sense, the foundational descriptions of photons, energy, and information are conceptually "nonmaterial" in nature.

In addition, contemporary physics posits the existence of dark energy (estimated to be approximately 68% of the known universe) plus dark matter (missing mass, estimated to be about 27%). This means that, given current assumptions, virtually 95% of the known universe is not only "dark" or imperceptible, but its existence can only be inferred at best.

Furthermore, there is the even deeper question regarding the empirical holograph-like nature of photonic information. This abstract, nonmaterial nature of light is often ignored by scientists and philosophers alike.

For example, it is relatively easy for a person with normal vision to see the North Star in a cloud-free night sky. The North Star has been estimated to be 434 light-years away from planet Earth.[3] In other words, patterns of photonic information and energy travelling from the North Star take about 434 Earth years to reach us. This means that when we see the North Star on a given night, what we are really seeing are patterns of light (or patterns of photonic information and energy) representing the state of the North Star as it was 434 years ago. In other words, we are not seeing the material North Star in present time, but, rather, we are seeing its nonmaterial dynamic, photonic, info-energy history as it has been "preserved" in the "vacuum" of space.

Although the North Star is estimated to have a radius of 35,400,000 kilometers, and by comparison, Earth's radius is only about 6,400 kilometers, the radius of the photonic patterns that reach us and pass through the lenses of our eyes is extraordinarily small: less than 0.000001 kilometers, or 1 millimeter.

Imagine that we are on a football field filled with people, with everyone looking at the North Star simultaneously. If there are 54,000 people on the field and they all had normal eyesight, 54,000 people could simultaneously report seeing the North Star.

What does this fact imply about the existence of multiple, nonmaterial North Star patterns existing simultaneously in space and time? The logical implication is that there are at least 108,000 tiny, nonmaterial North Star photonic patterns (one per eye, two eyes per person) simultaneously reaching the football field at any moment—and this, of course, is a gross underestimate of the actual total number. The nonmaterial nature of photonic light patterns strains our imagination as we try to comprehend this deep complexity.

When I teach this deep insight to students, I ask them to wonder how it is that, regardless of where they are sitting in a classroom consisting of a hundred students, they can see the complex patterns of light that define my hair, facial features, body features, clothing, and so on.

I then ask them to imagine how big this patterned information is inside their eyes (again, maybe 1 millimeter) compared to my actual height (170.2 centimeters).

Then I ask them how many tiny, nonmaterial, photonic Professor Schwartz's are in the room right now, given that each of the 100 students can see me simultaneously? They quickly realize there are at least 200—which is a profound underestimate.

After pondering this insight, I ask them to imagine that we are all standing outside, looking up at the sky, and that the nonmaterial photonic patterns representing each of us are being reflected up into space. I remind them that the existence of our individual, reflected, nonmaterial photonic patterns reaching outer space is a scientific fact established by spy satellites that detect these individual photonic patterns present in space, and therefore "see" us as individual beings.

I then ask them to try to imagine how many millions or billions (or some huge number) of an individual person's nonmaterial photonic patterns are spreading out, simultaneously and continuously, in the "vacuum" of space—that is, places in space having no material mass.

Finally, I ask them to imagine what happens to all this holistic, holograph-like nonmaterial information after a physical star dies, or a physical person dies. They realize that, in principle, it is still there. The individualized, nonmaterial, photonic info-energy patterns survive.

I shall end this discussion with a description of the discovery of the cosmic background radiation and the subsequent quantification of these ancient patterns of photons that are believed to have been created only a few hundred thousand years after the Big Bang. In other words, these historic photonic patterns, present everywhere in the universe, are presumed to be at least 13.7 billion years old, and they therefore have a kind of immortality.

It is as if the universe is a gigantic, nonmaterial photonic pattern (or info-energy) memory storage system. I raise the question, "Is it possible that the reason the universe appears to be expanding is to continue to make room for all this emerging historical information and energy?"

Other questions come to mind.

Is it possible that, in some deep sense, every "thing" that has ever existed continues its existence in some sustained, potentially evolving, info-energy way?

Is it possible that every "thing" that has ever existed subsequently has a nonmaterial, info-energy "afterlife," and that this survival potential is built into the very fabric of the universe, as expressed in the essence of light itself?

Does this mean that contemporary science allows for, and even predicts, the survival of consciousness after physical death, or what has been historically called soul?

Moreover, can we ponder what kind of Extraordinary Source Mind would be required to create such an extraordinary, photonic, quantum field-like universe?

My experience has been that students who "get" this nonmaterial reality insight never see the world in quite the same way again.

The distinguished physicist Richard Feynman said, "If you think you understand quantum mechanics, you don't understand quantum mechanics." Or, as Niels Bohr, one of the fathers of quantum mechanics, said, "If quantum mechanics hasn't profoundly shocked you, you haven't understood it yet."

A general, take-home message is: *As strange as the known properties of light and quantum mechanics are, they only hint at the inherent extraordinariness of the greater, nonmaterial reality.*

Part II: Insights into What Survival-of-Consciousness Research is Teaching Us about Nonmaterial Reality

Let us now turn to examples of what I have learned about nonmaterial reality as revealed through contemporary, life-after-death research.

Many areas of research contribute evidence that supports the survival-of-consciousness hypothesis. They include: (1) near-death research (for example, van Lommel et al. 2001); (2) past-lives research (for example, Tucker & Nidiffer 2014); (3) multi-blinded, controlled laboratory experiments with documented evidential mediums (for example, Sarraf et al. 2021); and (4) multi-blinded controlled laboratory experiments employing state-of-the-art, ultra-weak signal-detection hardware and software (for example, Schwartz 2021a). Reviews of this research, including its methodologies and statistics, can be found in the references.

However, in the spirit (no pun intended) of addressing insights learned about material and nonmaterial realities and their interactions, I will reveal four exemplary, extraordinary insights:

1. Specific and identifiable "deceased" individuals—whom we now refer to as "post-material persons" (PMPs)—not only deeply care about bridging nonmaterial and material realities but are also actively collaborating in this research (that is, they are post-material collaborators, or PMCs).

   I first revealed the identity of one of these hypothesized PMCs in my 2011 book, *The Sacred Promise: How Science is Discovering Spirit's Collaboration with Us in Our Daily Lives.*[4] The reason I titled this book *The Sacred Promise* was because the totality of the research provided compelling evidence indicating that these special, high-level PMCs have made three levels of profound promises for science and society:

Promise I: Specific PMCs have committed to actively collaborating in this research to help prove scientifically, once and for all, that the survival of consciousness is real.

Promise II: Specific PMCs have committed to fully complying with experimental protocols for the shared goal of scientifically establishing their participation, and we can depend on their participation. For example, they have agreed that if we, who are in the physical realm (the "B" Team), will show up in the laboratory for a given experiment, then they, who are in the realm of spirit (the "A" Team), will show up in the laboratory as well.

Promise III: If and when a practical PMP communication technology (referred to as the SoulPhone—see www.thesoulphonefoundation.org) becomes viable, they will continue to help humanity in their role as experienced and caring advisers and collaborators—if we (humanity) seek their assistance for the best and highest good.

The idea that material technology can be combined with specific individuals living in nonmaterial reality for the purpose of healing, restoration, and evolution is an inspiring, humbling, and sacred opportunity for humanity.

2. When evidential research mediums associated with the laboratory (such as Rhonda Schwartz, the chief selective evidential research medium) request members of the A Team to appear at spontaneous meetings, the PMCs see their responsibility as being "on call." Unless the PMCs indicate otherwise, the totality of the evidence is that they virtually always show up.

3. The evidence indicates that the ability of PMCs to "travel" from their nonmaterial reality to our material reality seems to operate at the "speed of thought." There are numerous instances where the behavior of PMCs, who exist in nonmaterial reality, is less constrained or obstructed than our behavior in material reality.

4. PMPs in general are typically perceived by evidential mediums as being

in physical-like form as they historically appeared on the Earth, in terms of physical size, shape, hairstyle and hair color, choice of clothing, and so on. Consequently, they can be recognized by their loved ones. Moreover, PMPs can produce physical-like effects on physical sensors—albeit, at greatly attenuated, ultra-weak levels—whether the sensors involve scientific precision optical instruments (such as silicon photomultipliers) or sensitive consumer devices (such as plasma globes). It is the physical-like nature of PMPs, at least as they appear on Earth, that promises to make it possible to invent electronic texting and video conferencing technologies to communicate with them (for example, see www.thesoulphonefoundation.org).

To illustrate some of the physical-like parallels between material and nonmaterial reality, I will briefly share two illustrations from controlled research where an evidential medium who has training as a portrait artist (and who prefers to remain anonymous) was requested to draw portraits of selected PMPs (reported in Schwartz 2021b).

One of the research paradigms that the late Susy Smith taught me, after she died, is what I have called the "double-deceased" paradigm (Schwartz 2011b). What this means is that one deceased person—for example, Susy Smith—is requested by a laboratory-tested evidential medium (R. Schwartz 2016) to bring another deceased person—for example, the distinguished quantum physicist David Bohm—to a second medium (previously tested or claimant[5]) under blinded conditions. In other words, in this instance, the claimant artist-medium (medium 2) was not told who Susy Smith (as invited by medium 1) is purportedly bringing to sit for a portrait.

Since the claimant artist-medium was open to being tested, and Susy Smith was apparently enthusiastic to demonstrate the medium's purported skills, we repeated this double-deceased experiment nine times with different hypothesized spirit collaborators who had been involved in previous afterlife research. Two portraits by claimant artist-mediums and photos of the PMPs who sat for her are shown below. As can be seen, the portraits closely match the photos of the secretly invited hypothesized spirit participants.

In **Figure 1**, the image on the left is a portrait of Susy Smith drawn by the medium; the picture on the right is a photograph I had taken of Susy a few years before she physically died. The resemblance of the portrait to the photograph is readily apparent.

Figure 1 Portrait by Medium Photo by Author

A second example is shown in **Figure 2**. The resemblance of the artist-medium's portrait to the photograph of the PMP (namely, Edgar Mitchell) is again apparent.

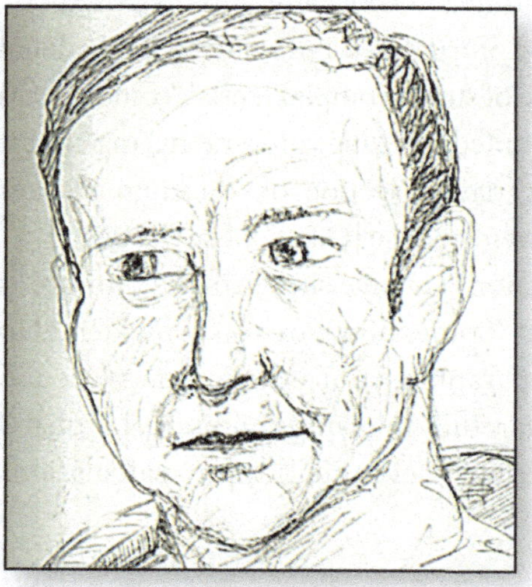

Figure 2 Portrait by Medium Photo by Author

I might mention that the claimant artist-medium (who was blind to the PMP's identity) spontaneously said, and I paraphrase slightly, that the unidentified spirit "thinks he's an astronaut," as if "the spirit must be a little crazy." The secret PMP we selected to be brought by Susy Smith to the medium's home was Edgar Mitchell.

Part III: Evidence for a Universal Source Mind Playing a Role in Human Life

Logic dictates that if a nonmaterial/greater reality exists and consists of Mind at its core, it follows that this nonmaterial Universal (Source) Mind should have the ability to influence the minds within it, and this includes human minds. Consequently, if the Universal Mind so chooses (with intelligence and intention), it can subtly influence our choices and decisions, and thereby create complex patterns of human activities, both in series (that is, sequentially over time) and in parallel (that is, simultaneously within time). Examples of complex, serial and parallel patterned events and processes, conceived and implemented by intelligent and intentional human beings, include (1) the composing of complex orchestrated music, and (2) the programming of complex, synchronized computer software.

Mathematical and statistical analyses confirm that such complex patterns of serial-parallel events cannot be achieved through random processes, especially when the constraints of physical time are considered. In my 2017 book, *Super Synchronicity: Where Science and Spirit Meet,* I provide arguably definitive evidence for the existence in real life of super-complex, serial "coincidences" or "synchronicities" that cannot be accounted for using conventional materialist explanations such as randomness, faulty data collection, or conventional external stimuli, such as physical forces or events per se (G.E. Schwartz 2017).

The totality of the serial coincidence evidence supports the humorous statement attributed to Yogi Berra that "Some things are too coincidental to be a coincidence"; or, as Susy Smith paraphrased it, "Some things are too coincidental to be accidental." When carefully applying conditional probability statistics to real-life, complex, serial synchronicities, my calculations reveal that the probability of such super synchronicities occurring by chance is "astronomically improbable." In other words, it would literally take longer than the age of the entire universe (roughly 13.7 billion years) for the documented sequence of events to have occurred by chance. Moreover, these astronomically improbable super synchronicities typically convey meaningful information of significance to our individual as well as collective lives.[6]

Part IV: Awaking to Nonmaterial Reality and the Primacy of Consciousness

When the controlled experiments involving mediumship and technology are considered as a whole, the totality of the findings strongly point to the nonmaterial reality of hypothesized PMPs.[7] However, skeptical scientists require definitive findings that can be independently replicated. Toward this end, we have been developing a paradigm to make it possible to conduct multi-experimenter, multi-center, multi-blinded, randomized control trials investigating hypothesized spirit presence and communication (Schwartz 2021a).

Large-scale, dual-center, validation experiments have been completed and submitted for publication in a mainstream journal. Below is a draft abstract of this paper (slightly edited for clarity). The term "hypothesized discarnate intelligences" was employed (rather than PMPs) to maintain consistency with terminology appropriate for a mainstream journal.

Abstract

The gold standard in biomedical research is the multi-center, multi-blinded, randomized control trial (RCT). An experimental randomized control method was devised to test for the presence of and communication with hypothesized discarnate intelligences. The method uses readily available laboratory equipment, automated data collection, and real-time analyses. A total of 1504 experimental sessions and 864 matched control sessions were run at two separate laboratories in the United States (in Arizona and Ohio). Using a controlled, binary, personal identification test, replicated findings were obtained independent of *investigators' beliefs about the presence of the discarnates, with experimental conditions associated with $p<.0001$, and control conditions associated with $p<.533$. The protocol involves an automated means of inviting discarnates to interact with high-voltage electrical plasma contained within a sphere. The methodology minimizes false-positive and false-negative outcomes. Independent RCT replications are underway employing skeptical experimenters. If the current findings are replicated, the evidence for the survival of human consciousness after death can be arguably interpreted as being definitive.*

If the multi-center studies produce replicate results as predicted, the simplest and most parsimonious explanation for the totality of the evidence, including genuine personal experiences, is the metaphysical reality of the presence of and communication with beings in a greater, nonmaterial reality.

Due to space and time constraints, my concluding remarks are limited to a few salient points.

First, ancient ideas about the nonmaterial nature of a greater reality are being re-examined and re-envisioned today. Contemporary scientific theory and research are lending ever-increasing support to statements such as, "Mind is the matrix of all matter," and "an unseen dimension [exists] behind the visible creation that gives it [the material universe] an intelligent design and that gives life purpose." The international conferences on "Science and God" exemplify this emerging trend.

Second, our current understanding of light, energy, and information not only illuminates inherent nonmaterial theory and research but also provides a practical "starting point" for approaching the relationship between material and nonmaterial reality. It includes foundational questions on the survival of consciousness after physical death and the existence of a pervasive Universal Mind that, in principle, could be playing a role in all aspects of the universe, from the micro to the macro dimensions of the cosmos.

Third, the integration of ancient wisdom with illustrative contemporary science (such as the survival of consciousness and synchronicity science) provides not only new inspiration and hope for humanity but also practical advice and collaboration from experienced and wise minds living in the greater reality, as well as from the One Mind itself (Dossey 2014).

Finally, in keeping with the concepts and evidence shared in this address, I raise this challenging question: Is it possible that beings in the non-material, greater reality, including the Source itself (for example, Laszlo 2020), are playing some sort of behind-the-scenes role in organizing these conferences on "Science and God"?[8] ◊

Endnotes

1. Some might prefer the word "resurrect," employing its more generic meaning, "to revive the practice, use, or memory of (something); bring new vigor to" (Oxford Languages).
2. For example, I received my PhD at Harvard University in psychology with a specialty in psychophysiology in 1971, and I was a mainstream, tenured professor of psychology and psychiatry at Yale University and director of the Yale Psychophysiology Center before I was recruited in 1988 to the University of Arizona.
3. Various studies have produced differing estimates of the distance of the North Star from Earth, ranging from 323 to 447.6 light-years. A study in the late 1990s with a European satellite gave a distance of 434 light-years, which has often been cited in the scientific literature.
4. The hypothesized PMC is Harry Houdini, the distinguished magician and revealer of fraudulent mediums.
5. That is, a person who claims to be an evidence-based medium.
6. Space and time preclude my sharing empirical data here, but substantial evidence has been reported elsewhere (G.E. Schwartz 2017).
7. And higher spiritual beings, including angels (for example, see Schwartz 2011b).
8. This is now becoming a logic-based and evidence-based testable scientific question.

References

Dossey, L. 2014. *One Mind: How Our Individual Mind Is Part of a Greater Consciousness and Why It Matters.* Carlsbad, CA: Hay House.

Laszlo, E. 2020. *Reconnecting to the Source: The New Science of Spiritual Experience, How It Can Change You, and How It Can Transform the World.* New York, NY: St. Martin's Press.

Sarraf, M., M.A. Woodley, and P. Tressoldi. 2021. "Anomalous Information Reception by Mediums: A Meta-Analysis of Scientific Evidence." *EXPLORE: The Journal of Science and Healing* 17 (5): 396–402.

Schwartz, G.E. 2010. "Possible Application of silicon Photomultiplier Technology to Detect the Presence of Spirit and Intention: Three Proof-of-Concept Experiments." *EXPLORE: The Journal of Science and Healing* 6 (3): 166–171.

———. 2011a. "Photonic Measurement of Apparent Presence of Spirit Using a Computer Automated System." *EXPLORE: The Journal of Science and Healing* 7 (2): 100–109.

———. 2011b. *The Sacred Promise: How Science Is Discovering Spirit's Collaboration with Us in Our Daily Lives.* Hillsboro, OR: Atria Books/Beyond Words.

———. 2017. *Super Synchronicity: Where Science and Spirit Meet.* Cardiff-by-the-Sea, CA: Waterside Digital Press.

———. 2021a. "A Computer-Automated, Multi-Center, Multi-Blinded, Randomized Control Trial Evaluating Hypothesized Spirit Presence and Communication." *EXPLORE: The Journal of Science and Healing* 17 (4): 351–359.

———. 2021b. "Contemporary Evidence for Communication with the Departed and the Sacred." In *Human Interaction with the Divine, the Sacred, and the Deceased: Psychological, Scientific, and Theological Perspectives,* edited by Thomas Plante and Gary E. Schwartz. New York, NY: Routledge.

Schwartz, R.E. 2017. *Love Eternal: Extraordinary Personal and Scientific Evidence for Life after Death* (2nd ed.). Vancouver, BC: Param Media.

Schwartz, S.A., M.H. Woollacott, and G.E. Schwartz, Eds. 2020. *Is Consciousness Primary?* Tucson, AZ: AAPS Press.

Tucker, J.B., and F.D. Nidiffer. 2014. Psychological Evaluation of American Children Who Report Memories of Previous Lives. *Journal of Scientific Exploration* 28 (4): 583–594.

van Lommel, P., R. van Wees, V. Meyers, and I. Elfferich. 2001. "Near-Death Experience in Survivors of Cardiac Arrest: A Prospective Study in the Netherlands." *The Lancet* 358 (9298): 2039–2045.

Acknowledgments

I thank Mark Pitstick, MA, DC, Darlene Martin, MA, Rhonda Schwartz, MA, and Dinshaw Dadachanji, PhD, for their editorial comments.

About the Speaker

- Professor of Psychology, Medicine, Neurology, Psychiatry and Surgery, The University of Arizona, Tucson, USA
- Director, Laboratory for Advances in Consciousness and Health
- Founding President, Academy for the Advancement of Postmaterialist Sciences
- Areas of expertise include: alternative medicine and healing; consciousness and life after death; integrative medicine; psychophysiology; spiritual psychology
- Books authored and coauthored include: *The G.O.D. Experiments: How Science Is Discovering God in Everything, Including Us* (2007); *The Sacred Promise: How Science Is Discovering Spirit's Collaboration with Us in Our Daily Lives* (2011); *Super Synchronicity: Where Science and Spirit Meet* (2017); *Greater Reality Living: Integrating the Evidence for Eternal Consciousness* (2018); *Extraordinary Claims Require Extraordinary Evidence: The Science and Ethics of Truth Seeking and Truth Abuse* (2021)
- PhD in Psychology, (1971, Harvard University)

ICSG II

SESSION 1

Unraveling the Secrets of the Universe: The Progress of Scientific History

About Session 1

Unraveling the Secrets of the Universe: The Progress of Scientific History

The journey of science to investigate the natural world began by probing the things we can perceive with our five senses. These studies led us to recognize the existence of certain invisible, intangible fields, such as the electromagnetic field and the gravitational field. At the turn of the twentieth century, when examining the fundamental structure of matter, scientists were compelled to theorize the existence of additional intangible fields and forces–the weak and strong nuclear forces within atoms–and to accept the revolutionary ideas of quantum physics. Next, astrophysicists studying distant galaxies realized the need to postulate that most of the universe is constituted of mysterious dark matter and dark energy. In recent years, scientific concepts about the nature of the universe are continuing to evolve dramatically, prompted by the study of consciousness. Thus, science has been going through periods that we may call classical science, then modern science, and now, "frontier science" or "new paradigm science." This session looks into evidences of the nonlocal effects of the mind, challenging materialist thought and pointing to the start of the postmaterialist era. ◊

About the Chair, Dr. Cassandra Vieten

- Executive Director, John W. Brick Mental Health Foundation, Timonium, Maryland
- Director of Research, Arthur C. Clarke Center for Human Imagination at the University of California, San Diego
- Senior Fellow and past President (2013–2019), Institute of Noetic Sciences, Petaluma, California
- Clinical psychologist, mind-body medicine researcher
- Author of numerous scientific articles; author of *Mindful Motherhood;* coauthor of *Living Deeply* and *Spiritual and Religious Competencies in Clinical Practice*
- PhD in Clinical Psychology (1998, California Institute of Integral Studies)

Session 1 | Presentation One

The Flow of Science from Tangible to Intangible Dimensions: Some Implications from Research into Anomalous Healing

Professor William Bengston
Statistics and Research Methods,
St. Joseph's College, New York, USA

A phenomenon is considered scientifically "anomalous" if it does not fit into the mainstream understanding of how the world is thought to work. It does not mean that the phenomenon is not real; it just means that it does not make sense to us. The problem may lie not with the facts of the phenomenon but in our lack of theoretical understanding. Anomalous phenomena, if demonstrated to be valid and reliable, can suggest to us that it might be time to rethink our theoretical understanding (Kuhn 1970).

About the Speaker

- Professor of Statistics and Research Methods, St. Joseph's College, Patchogue, New York
- President, Society for Scientific Exploration (an international scientific society in 27 countries that studies scientific anomalies)
- Author of *The Energy Cure*; numerous peer-reviewed publications
- PhD in Sociology, Fordham University

Take the case of healing. Virtually all recorded societies report that certain individuals appear to have the ability to heal. Oftentimes this healing has been associated with spiritual disciplines of one sort or another, and the healers themselves have sometimes been accorded a special status within the culture. Healers have utilized various methods of practice, including laying-on of hands, prayer, and induced altered states of consciousness, to name a few. The father of Western medicine, Hippocrates, referred to this healing as "the force which flows from many people's hands" (Schiegl 1983).

The foundation of modern Western healing research can be traced to the pioneering work of the late biologist Bernard Grad at McGill University. In carefully controlled experiments, Grad found that selected healers could influence the germination of plant seeds, the growth rate of plants, and the curing of seeds that had been shocked by saline solution. In addition, he was able to measure the ability of healers to reduce goiter and stimulate wound healing in mice (Grad, Cadoret, and Paul 1961; Grad 1963, 1964). Since Grad's initial work, there have been innumerable preclinical studies of healing, sometimes categorized by the target of the intended healing. Benor, for example, discusses healing action on enzymes, cells in the laboratory, fungi/yeasts, bacteria, plants, single-cell organisms, and animals that have been subjected to controlled study (Benor 1992).

For a very long time after Grad's pioneering work, researchers found themselves on the defensive. That is, they were stuck in a dialogue with supposed "skeptics," repeatedly demonstrating rigorous scientifically controlled data while being met by scorn. That dialogue consisted of mutually antagonistic groups talking past each other, as healing researchers led with rigorous data and skeptics (who are oftentimes "debunkers" in disguise) countered with theoretical impossibility.

Fortunately, in recent years, the proliferation of healing studies has greatly risen. At present, there are several peer-reviewed journals devoted exclusively to the burgeoning field of complementary and alternative medicine (CAM), publishing both preclinical and controlled clinical studies of healing of a wide variety of conditions (for example, the *Journal of Alter-*

native and Complementary Medicine; Alternative Therapies in Health and Medicine; Explore: The Journal of Science and Healing). In addition, there are an increasing number of peer-reviewed journals that are not focused exclusively on CAM but are open to publishing controlled studies in these areas (for example, the *Journal of Scientific Exploration*).

Researchers have spent an inordinate amount of time and effort trying to demonstrate the "fact" of healing to the so-called skeptical community, and too little effort looking for what might be termed the "secondary correlates" of healing.

For over 35 years, I have been researching anomalous healing using a healing method I developed (Bengston 2007, 2010), studying both *in vitro* and *in vivo* models. My research agenda has looked at some parameters of healing such as distance and dose; the physiological correlates of healing; and more recently, an attempt to reverse engineer the healing effect to make it a scalable and reliable conventional treatment.

Eighteen *in vivo* cancer experiments on mice have been conducted, using standard models of mammary adenocarcinoma, methylcholanthrene-induced sarcomas, naturally occurring oncogenic tumors, immune-deficient nude mice, and on an extremely aggressive cancer model (Bengston and Krinsley 2000; Bengston and Moga 2007; Bengston 2004, 2017; Moga and Bengston 2010; Beseme 2018). These have been performed at eight different institutions, including six medical schools. In addition, innumerable *in vitro* experiments have been performed on human leukemia and breast cancer cells, to name but a few.

The experimental protocol used in all these experiments has been to take a "standard" mouse or cell model with a long history of conventional empirical research that has a known and predictable outcome (Bengston and Krinsley 2000; Bengston and Moga 2007; Bengston 2012; Lerner et al. 1966), and to introduce the variable of "healing with intent" using a healing technique that I helped develop (Bengston 2007). Volunteer healers, who have included both students and faculty, had all been pre-screened to be without any experience in alternative healing, nor were they in any way "believers" in the validity of alternative healing. Their healing experiences in my experiments were the first of any kind for them. Variations in many parameters have been examined, such as distance and dose and frequency of treatment, as well as the compilation of the subjective experiences of the volunteer healers. Additionally, human physiological correlates have been examined using electroencephalograms (EEGs) at a private lab (Hendricks et al. 2010) and functional magnetic resonance imaging (fMRI) (Bengston 2017) independently carried out at two medical schools. Moreover, we have measured the changes in the physical space adjacent to the healings (Moga and Bengston 2010).

A summary of the results of these experiments includes the following:

- Demonstration of a reliable, full-lifespan cure of cancer in experimental mice, including an apparent immunity to reinjection of the same cancer (Bengston and Krinsley 2000; Bengston and Moga 2007).
- Dose response to healing, with some minimum amount of healing time being necessary to effect a cure. Interestingly, the only predictor of the aggregate speed of cure is the number of mice in an experiment, the quicker cures being associated with more mice being treated (Bengston and Moga 2007).
- Healing apparently proceeds in nonlinear fashion, with sudden bursts of healing that resemble "phase transitions" (Bengston and Krinsley 2000).
- A measurable "resonant bond" between healer and healee is fluid, with successful healing being associated with "connection," and healing failure being associated with "disconnection" (Bengston and Moga 2007; Bengston 2012)."
- Healing shows no relationship to distance and appears to be fundamentally about "information" rather than "energy" (Moga and Bengston 2010).
- Healing can be demonstrated to be "storable" in both biological systems and physical systems (Beseme et al. 2018, 2020).

Subsequent discussion will move in the direction of reinforcing that (1) healing is not related to conscious awareness but rather is more akin to an autonomic biological response to need (Bengston 2017); (2) conscious intention and attention are quite dissimilar, with only fleeting intention being the necessary component of healing (Bengston and Krinsley 2000; Bengston 2010) and (3) intention itself may be *storable* (Beseme et al. 2018).

The end result will be that *need*, rather than conscious awareness, may be the driving force behind a great deal of what can be considered anomalous healing (Bengston 2017).

How consciousness fits into all of this will be explored but not fully explained. While anything approaching a comprehensive definition of consciousness remains elusive, I will suggest that a more useful conceptual model flows from the distinction between *awareness* and *intention*, the latter being the dominant partner and the former being optional at best.

The *in vivo* experimental model

In the basic experimental *in vivo* protocol, mice are subcutaneously injected with a known dose of cancer sufficient to guarantee death within a specified interval. Healing treatment of the mice generally involves the volunteer healer placing his or her hands on the outside of the cages and practicing the healing technique for a specific duration (Bengston 2007, 2010). Individual treatment length, number of treatments, and number of mice per treatment have all been part of various experimental protocols, including other variables such as distance of hands from the cages, extending to thousands of miles.

Those mice that have been treated by the healing with intent techniques typically develop an encrusted blackened area on the surface of the tumor (**Figure 1**), followed by tumor ulceration (**Figure 2**), and then implosion to full lifespan cure. At all stages of remission there are viable cancer cells in the mouse. When full cure is achieved, the mouse is completely free of cancer and is further apparently immune to subsequent injections of the same cancer for its entire life.

Sudden shifts, analogous to *phase transitions,* have been the pattern in

Figure 1 An experimental mouse (identified as "A-4") develops a tumor, seen here on day 28 after injection with cancer cells (Bengston and Krinsley 2000).

all experiments, whether *in vivo* or *in vitro*. That is, in early stages of healing treatment, mice (and cell cultures) show no apparent effect of the healing intention, until suddenly there are nonlinear, dramatic shifts in tumors (*in vivo* models) or cell growth (*in vitro* models).

Healing and the sense of connection: EEG data

In addition to the anomalous healing in and of itself, there is also apparently an anomalous connection that can occur between subjects. Synchronizing EEGs between the healer and the subject indicate that the healer's EEG data shows harmonic frequency coupling across the spectra, followed by frequency entrainment effects with the healee (**Figure 3**). These results suggest the presence of a connection between the healer and healee (Hendricks et al. 2010). Of central importance, in addition to the apparent connection established at a distance, is that neither the healer nor healee is consciously aware of the connection.

The tumor in an experimental mouse (identified as "A-3") undergoes ulceration, seen here on day 35 after injection with cancer cells (Bengston and Krinsley 2000). **Figure 2**

7.81 Hz spindle with harmonics at 15.63 and 23.44 Hz., bispectra and spectra

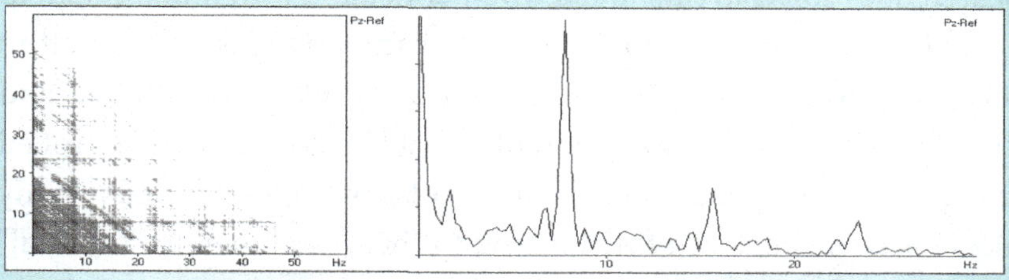

On the left is the bispectral display showing the cross spectral interaction of the 7.81 Hz EEG activity and the other mathematically related harmonics, seen as a "checkering" of the display. On the right are the spectral magnitude peaks, seen with the 7.81 Hz primary peak, and then the progressively smaller magnitude responses as the doubling and tripling harmonic peaks are seen.

The plot on the left displays all the cross-correlations between the brains of the healer and the healee separated by some distance. The graph on the right displays the dominant frequencies from brain measurements (Hendricks et al. 2010). **Figure 3**

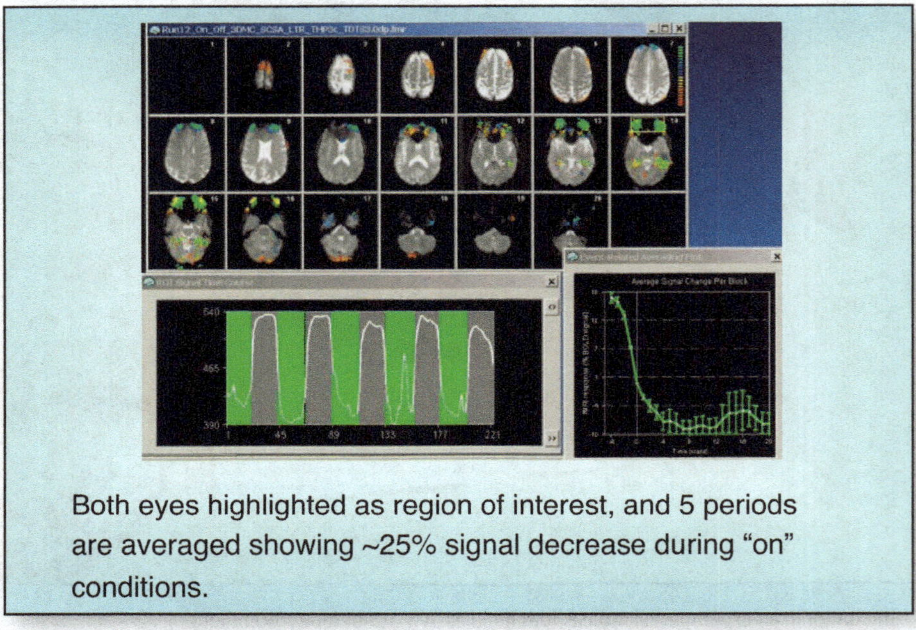

Figure 4 This figure illustrates the ability to turn healing intention "on" and "off" (Bengston 2017).

Healing and the sense of connection: fMRI data

Research questions have included whether healing with intent might have a specific location in the brain. In order to gather data using fMRI, it would be necessary to produce a "toggling" of healing intention into "on" and "off" states. To my skeptical amazement, this could be done under controlled conditions. At two independent medical schools, an exploratory pilot study had healers inside an enclosed fMRI intend to "heal" and then to "not heal" during 45-second cycles to see if healing intention can be toggled (Bengston 2017) (**Figure 4**). Apparently, it can.

Changes in physical space associated with healing

A few previous studies have reported magnetic field changes during bioenergy healing. In one experiment, Margaret Moga and I examined magnetic field activity during hands-on healing and distant healing of mice

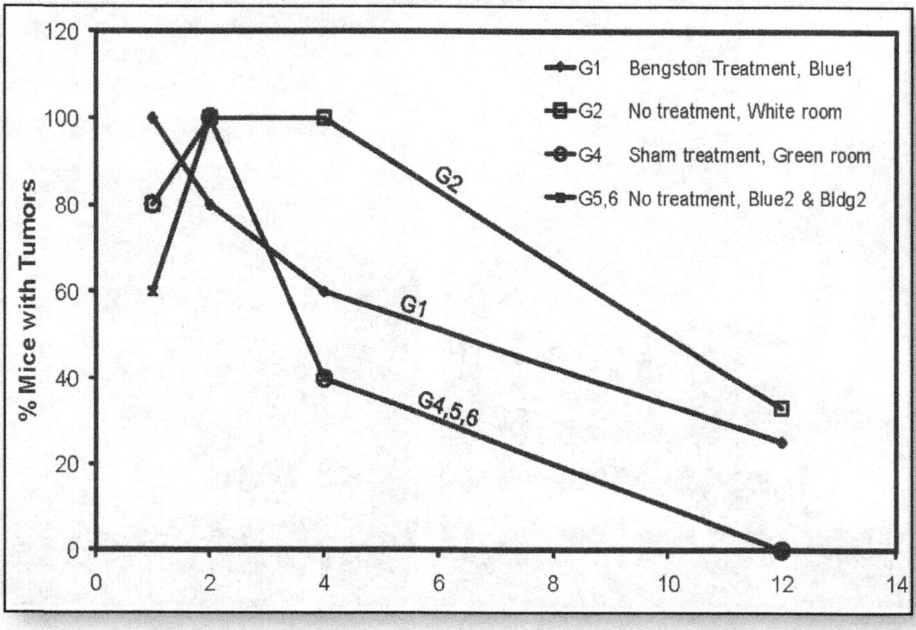

This figure shows that there are likely physical correlates to healing that can be measured by conventional devices used for environmental measurements (Moga and Bengston 2010). **Figure 5**

with induced tumors (**Figure 5**). During healing sessions, we observed distinct magnetic field oscillations adjacent to the mice cages, which were similar in appearance to those reported by Zimmerman (1985). Evidence to date seems to suggest that bioenergy healing may have physical correlates detectable with DC gaussmeters (Moga and Bengston 2010).

Anecdotally, we have also observed similar anomalous output from random number generators placed in the healing environment. That is, when the mice still have cancer, and only during healing treatments, random number generators placed inside the animal rooms produce anomalous data consisting of statistically significant deviations from chance output. *Both the anomalous magnetic field output and the random number output were suggestive of information being inserted into a physical system.* It is likely that we are capturing physical correlates to healing rather than the healing information itself.

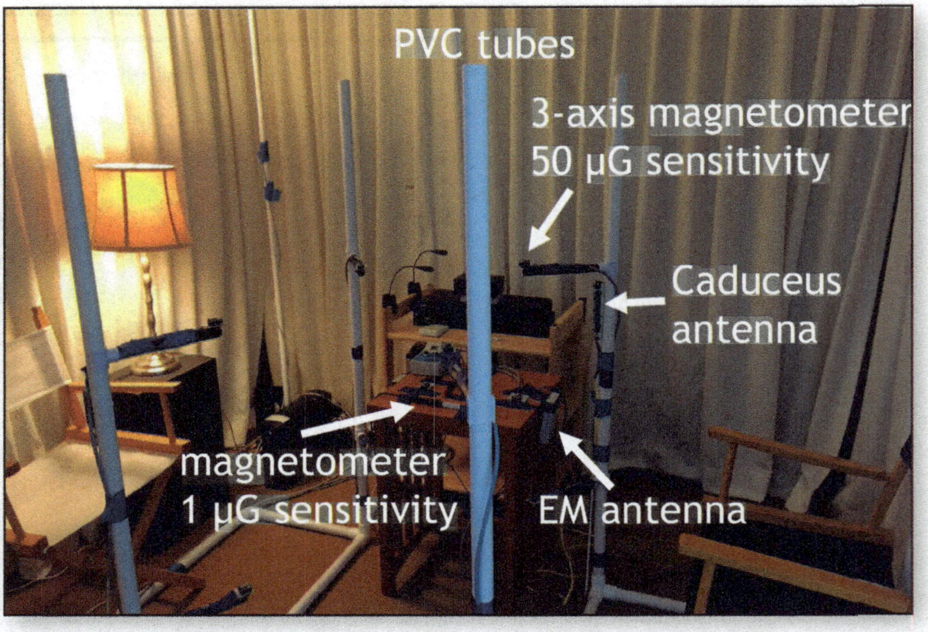

Figure 6 Four types of sensors were used to detect and record magnetic and electromagnetic signals when cotton was "treated" with Prof. Bengston's healing technique (Beseme et al. 2020).

Current experiments: Capturing and recording healing information

In some recent experiments, we have tried to record the healing information seemingly produced in a healing session, and to play back that recording to reproduce the healing effect without the healer (Beseme et al. 2018, 2020).

Three people practiced my healing method by "treating" cotton inside a solid steel, double-walled, electromagnetically shielded chamber for five minutes. As the treatment of cotton took place, magnetic and electromagnetic signals were recorded using four types of sensors (**Figure 6**).

Cancer cells *in vitro* were exposed to that recording, and 167 genes were tested. When cells were exposed to the recording at various time points, 68 genes showed statistically significant fold changes (Beseme et al. 2018). Most of these genes belonged to the cancer pathway assay, suggesting a possible anticancer effect from the recording.

The transcriptional changes indicate that the recording has a biological effect on cancer cells both *in vitro* (Beseme et al. 2018) and *in vivo* (Beseme et al. 2020). Some implications of these experiments suggest a possible technology of healing.

In order to compare the relative strength of a "live," hands-on treatment to the recording, we investigated whether the expression of selected genes was differentially affected. Interestingly, the transcriptional changes associated with the hands-on method occurred earlier than with the recording. The amplitude of the change was much greater when using the hands-on method than the recording. These two observations combined suggested that some *information* may be lost in the process of recording and delivering the healing. At this point it is unknown whether the playback of multiple recordings might in fact compensate for the apparent weaker signal.

Among the more interesting anecdotal observations was that "sensitives" exposed to the healing recording reported that even after the recording was turned off, they could still "feel" the recording as if it was continuing to play. So, we placed cancer cells inside the incubator three days *after the recording had been turned off,* and we found that the transcriptional changes in the cells continued to occur. This was similar to previous findings by William Tiller on what he termed "conditioned space" (Tiller 1997).

Discussion and conclusions

It should be apparent that these data challenge many scientific models about the way we think the world works, and certainly that the "classic" materialist model is insufficient. Any "postmaterialist" model that may emerge in the future must grapple with some significant challenges.

First, the theme of "need" consistently occurs. Cells that have a healing need will transcriptionally respond to healing with intent, whether that healing source comes directly from the hands of a healer, or healing apparently stored in substances such as water, cell medium, or cotton. Cells that have no healing need exhibit no anomalous transcriptional changes when offered healing with intent. Thus, at a minimum, it can be posited that biological need is a crucial component in healing, and it may be the healee that instigates the healing effect.

Second, conscious awareness on the part of either the healer or healee is not likely to be central to produce a healing effect. The extent or quality of consciousness on the part of mice or cells may be debatable, but there is little question that they are unlike anything that parallels human consciousness. Yet mice in need "know" to move proximate to a healing source; cells in need do likewise. Can there be serious doubt that these responses to healing are natural biological responses?

The volunteer healer logs vary widely in the extent to which they were consciously aware of anything associated with healing. Some occasionally felt some sort of "connection" with their mice; some felt nothing at all, the latter to the point that they seemed not to understand the question when asked to comment about feelings associated with healing. In multiple experiments, there has been no association between healing efficacy and subjective states of connection. Thus, *it may be posited that a conscious awareness of healing or connection may be unnecessary for healing to take place.*

At the same time, both EEG and fMRI data clearly indicate that some sort of biological connection does take place. For one, at least in the case of healing humans, healer and healee go into harmonic brain phase locking, and do so without any necessary conscious awareness that the healing phenomenon might be taking place.

Yet, while there can be high confidence that conscious *awareness* on the part of either the healer or healee is optional at best, the role of *intention* on the part of the healer becomes more problematic. That is, the simple act of putting hands around a cage, or attempting to "charge" materials for a healing experiment, signifies intention of some sort. That intention may be fleeting, and certainly separate from anything approaching either belief or sustained awareness. But if action is taken in order to produce or to test healing of any sort, there must be intention. This intention is akin to that expended for many forms of action. One "intends" to walk down the street, but there is nothing approaching "belief" or sustained "attention." Indeed, attentive walking will diminish efficacy. There must have been some intention to begin the walking, but the activity is driven not by sustained effort or attention but by a letting go. The mastery of many skills, whether walking or healing, likely involves the transition from "mindful" attention to relatively "mindless" fleeting intention.

Connection can be seen as an autonomic response to need. Consider that in one fMRI protocol, blinded envelopes with pictures and hair samples of animals placed onto the palm produced significant brain response in the healer if *need,* defined as pictures of cancerous animals, was present in those envelopes. These responses were biologically similar to the brain changes that occurred if the healer intentionally attempted to heal. If the envelopes placed into the palms of volunteer healers did not contain need, then no brain changes ensued. Again, there was no conscious awareness of whether any envelopes did or did not have any pictures of animals in need.

The commonly found association between certain states of consciousness and healing, often associated with being "spiritual," likely has the temporal sequence inverted. Instead of a "spiritual" sense of connection being necessary in order to produce healing, the data indicate that healing occurs more as an autonomic response to biological need, and the subjective sense of spiritual connection is an optional consequence of that need. Since more subjectively sensitive individuals are more likely to be drawn to healing, the mistaken association can be made that this sensitivity is the source of healing. It turns out that people who are less subjectively sensitive can heal just as well, without ever experiencing connection. *Conscious awareness of spiritual connection is optional.*

That healing intention can apparently be stored in materials and later be used to stimulate healing effects is extremely suggestive that consciousness may have an associative technology. The apparent storage of healing in water, cell medium, cotton, and even in a recording, which can produce a future healing effect when need is present, begs inquiry into future studies that may help to unravel some of the mysteries of healing. And there is the additional possibility that this storage ability might be able to make healing more conventional and scalable.

Finally, the lack of necessity of awareness of spiritual connection on either the part of the healer or healee makes it likely that healing does not conform to models of psychokinesis which support conscious intention as the operative agent. That is, healing outcome is not "willed" in the way that operators can bring about intended alterations in, say, the theoretical output of random number generators. Volunteer healers may have "intended" healing in that they went through some training in a healing technique and placed their hands around cages, but there are no cases of these healers having healing follow their desires, either that the tumors never form or that they shrink immediately in response to the healing effort. Indeed, initial experiments proceeded under the assumption that if healing were to work, then mice that were treated shortly after being injected with cancer would avoid tumor growth altogether. In all cases, regardless of type of cancer, and regardless of how soon after injection treatment began, tumors grew, sometimes very large, before the process of ulceration and implosion commenced. The volunteer healers were successful in the outcome; they were upset and concerned when their mice developed tumors. Certainly, the pattern and stages of healing does not conform to the wishes of the healers.

The data output from the experiments, and the experiences of the healers, do not conform to anything like a direct psychokinetic (PK) effect. Instead, there is merit to thinking of healing as a non-directed outcome, similar to that proposed by Robert Jahn and Brenda Dunne in their "M5 model" for explaining their consciousness-related anomalies with random event generators and remote perception studies (at the Princeton Engineering Anomalies Research laboratory). In that model, the notion is proposed that the conscious mind might connect to the tangible physical world not directly but by way of a circuitous route involving unconscious processes and intangible physical mechanisms. Further speculation involves a timeless and spaceless "Source" in which the unconscious and intangible merge.

While a full examination of the application of the M5 model to healing is beyond the scope of this discussion, the actual healing techniques used in these experiments, as well as the subjective experiences of a selection of volunteer healers are remarkably consistent with this model. This includes speculative discussion of the Source, which is directly discussed elsewhere (Jahn and Dunne 2001, 2011). A full explication of the usefulness of the model to an understanding of healing, however, would actually and controversially minimize the importance of the conscious mind.

The major question, the elucidation of the healing information, remains hidden, even as one discovers biological and physical correlates. Changes in the brain, changes to the physical space of the healing environment, and so forth should be seen as secondary correlates to the healing. While interesting, these secondary correlates are not likely to be capable of unraveling the primacy of the healing information.

This is suggestive of the important distinction between "energy" and "information." The biological and physical correlates that were observed are likely energetic signatures, which are secondary correlates to the actual healing effect. Healing itself, however, can be seen to be "informational" in nature, presumably transmitted to the healee through an energetic process not yet understood. The analogy might be to a radio signal, which can be seen as an information packet riding on a power source. Here, healing is the information or "message" packet, and the power source remains a mystery. Since healing apparently does not diminish with distance, that power source must be global in reach.

So many questions beg to be addressed, to the point of being overwhelming. We have obviously just begun to scratch the surface in our understanding of these phenomena. As the twentieth century found the equivalence between matter and energy, in the twenty-first century perhaps we are on the verge of finding an equivalence between energy and information, to be followed by undreamt-of technologies and applications.

We are clearly on the verge of yet another scientific revolution. ◊

References

Bengston, W. 2004. "Methodological difficulties involving control groups in healing research." *Journal of Alternative and Complementary Medicine* (renamed *Journal of Integrative and Complementary Medicine*) 10 (2): 227–30.

——, 2007. "Commentary: A method used to train skeptical volunteers to heal in an experimental setting." *Journal of Alternative and Complementary Medicine* (renamed *Journal of Integrative and Complementary Medicine*) 13 (3): 329–32.

——, 2010. *Hands-On Healing: A Training Course in the Energy Cure* (Audio CD). Sounds True.

——, 2012. "Spirituality, Connection, and Healing with Intent: Reflections on Cancer Experiments on Laboratory Mice." In *Science and Spirituality,* edited by Lisa Miller. New York, NY: Oxford University Press.

——, 2017. "Some Reflections on Consciousness, Intention, and Healing," In *Being and Biology: Is Consciousness the Life Force?* Edited by B. Dunne and R. Jahn. Princeton, NJ: ICRL Press.

——, and D. Krinsley. 2000. "The Effect of the 'Laying On of Hands' on Transplanted Breast Cancer in Mice." *Journal of Scientific Exploration* 14 (3): 353–64.

——, and M. Moga. 2007. "Resonance, Placebo Effects, and Type II Errors: Some Implications from Healing Research for Experimental Methods." *Journal of Alternative and Complementary Medicine* (renamed *Journal of Integrative and Complementary Medicine)* 13 (3): 317–27.

Benor, D. 1992. *Healing Research: Holistic Energy Medicine and Spirituality.* Vol. 1. Munich, Germany: Helix Editions.

Beseme, S., W. Bengston, D. Radin, M. Turner, and J. McMichael. Jul–Sep 2018. "Transcriptional Changes in Cancer Cells Induced by Exposure to a Healing Method." *Dose-Response* 16 (3): 1559325818782843.

——, L. Fast, W. Bengston, M. Turner, D. Radin, and J. McMichael. Jan–Mar 2020. "Effects Induced in vivo by Exposure to Magnetic Signals Derived from a Healing Technique." *Dose-Response* 18 (1): 1559325820907741.

Grad, B. 1963. "A Telekinetic Effect on Plant Growth." *International Journal of Parapsychology* 5: 117–33.

——. 1964. "A Telekinetic Effect on Plant Growth. II. Experiments Involving Treatment of Saline in Stoppered Bottles." *International Journal of Parapsychology* 6: 473–98.

———, R. Cadoret, and G. Paul. 1961. "The Influence of an Unorthodox Method of Treatment on Wound Healing in Mice." *International Journal of Parapsychology* 3: 5–24.

Hendricks, L., W. Bengston, and J. Gunkelman. 2010. "The Healing Connection: EEG Harmonics, Entrainment, and Schumann's Resonances." *Journal of Scientific Exploration* 24(4): 655–66.

Jahn, R., and B. Dunne. 2001. "A Modular Model of Mind/Matter Manifestations (M5)." *Journal of Scientific Exploration* 15(3): 299–329.

———, and B. Dunne. 2011. *Consciousness and the Source of Reality: The PEAR Odyssey.* Princeton, NJ: ICRL Press.

Kuhn, T. 1970. *The Structure of Scientific Revolutions.* Chicago: University of Chicago Press.

Lerner, L., A. Bianchi, and M. Dzelzkalns. 1966. "Effect of Hydroxyurea on Growth of a Transplantable Mouse Mammary Adenocarcinoma." *Cancer Research* 26 (11): 2297–2300.

Moga, M., and W. Bengston. 2010. "Anomalous Magnetic Field Activity During a Bioenergy Healing Experiment." *Journal of Scientific Exploration* 24 (3): 397–410.

Schiegl, H. 1983. *Healing Magnetism.* Freiburg, Germany: Hermann Verlag KG.

Sheldrake, R. 1981. *A New Science of Life: The Hypothesis of Formative Causation.* New York: J.P. Tarcher.

———, 1995. *The Presence of the Past: Morphic Resonance and the Habits of Nature.* Rochester, VT: Park Street Press.

Tiller, W. 1997. *Science and Human Transformation: Subtle Energies, Intentionality and Consciousness.* Walnut Creek, CA: Pavior.

Zimmerman, J. 1985. "New Technologies Detect Effects of Healing Hands." *Brain/Mind Bulletin* 10 (3).

Session 1 | First Commentary
On the Presentation by Professor William Bengston

The Flow of Science from Tangible to Intangible Dimensions

Professor Margaret Moga
School of Medicine, Indiana University, USA

Scientific studies have discovered unique "bioemissions" from healers' hands and/or bodies. A leader in the field of energy healing, Dr. Bengston has proposed "reverse-engineering" the healer's bioemissions and using the healer's vibrations as the foundation for a new medicine based on natural healing frequencies (**Figure 1**).

In 2005, my laboratory in Indiana had the pleasure of working with Dr. Bengston in a set of experiments involving mice with tumors (Moga and Bengston 2010). We witnessed firsthand the tumor remission sequence as described by Dr. Bengston in response to hands-on and distant healing. The tumors grew rapidly but then suddenly began to disappear very quickly.

About the Speaker

- Associate Professor of Anatomy, Cell Biology & Physiology, School of Medicine, Indiana University
- Current research focus: developing scientific measures of "energy healing" and other bioenergetic phenomena; studying the exceptional experiences of healers
- Council member, Society for Scientific Exploration, promoting the Frontier Sciences
- PhD in Neuroanatomy; trained as a practitioner in Healing Touch, an energy-based biofield therapy

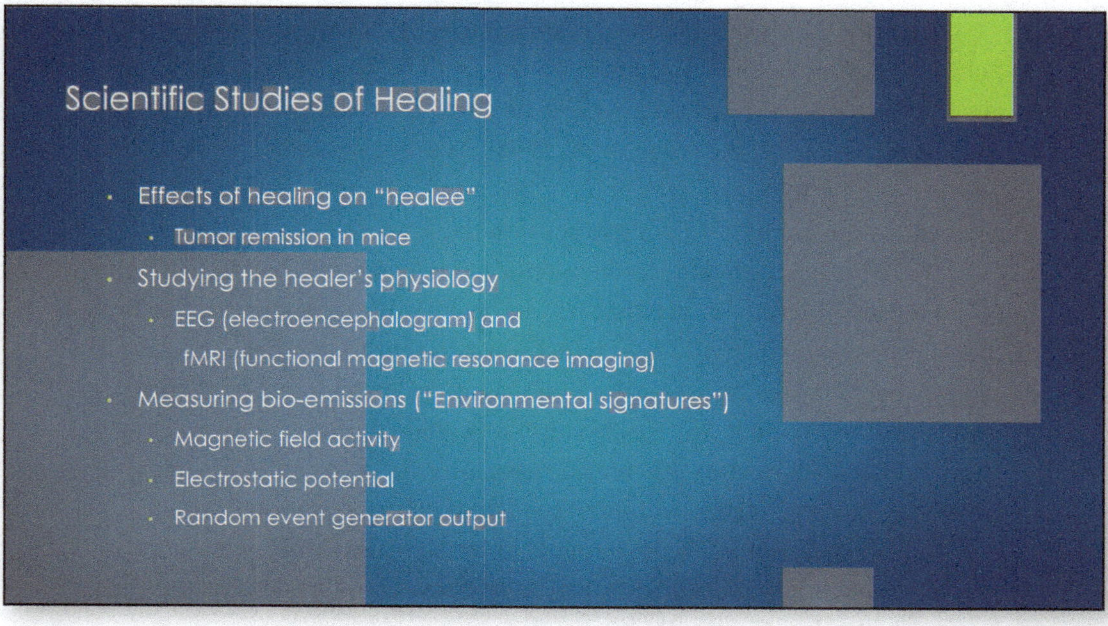

Figure 1

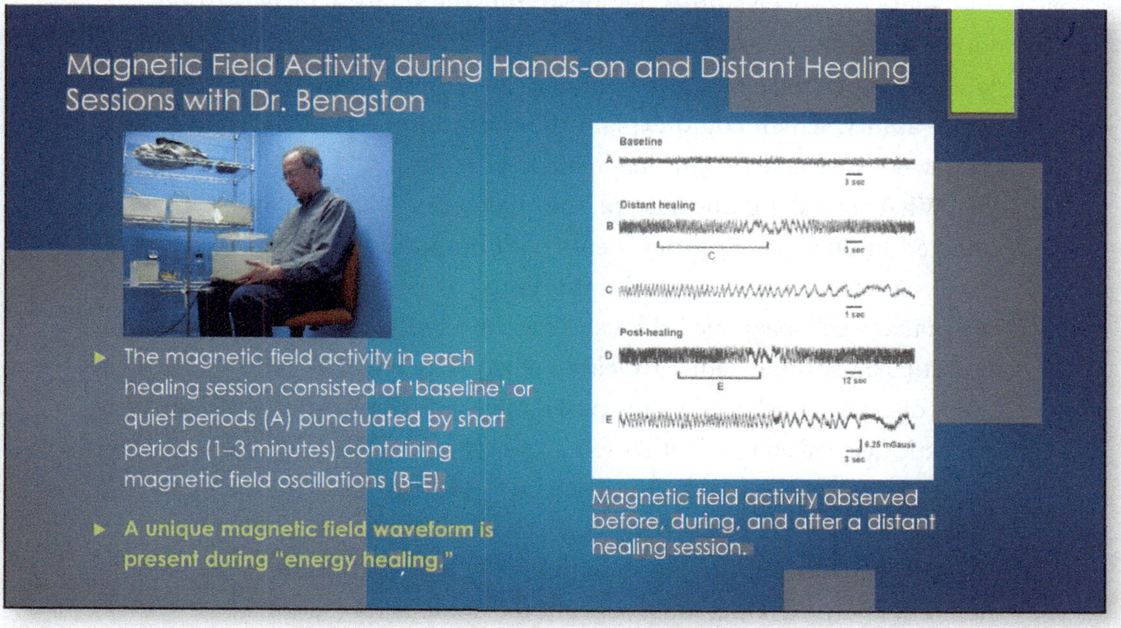

Figure 2

Previous studies had demonstrated electric field surges and magnetic field oscillations in close proximity to the healers' bodies, suggesting that electromagnetic field variations might be part of the energy healing mechanism (Seto et al. 1992; Green et al. 1991). We were particularly eager to duplicate the findings of Seto et al., who noted magnetic field oscillations adjacent to the hands of Qigong practitioners.

We did magnetic field recordings close to Dr. Bengston's hands, head, and body, and also adjacent to the mice he was healing. We were surprised to find that the strongest magnetic field signal was adjacent to the mice and mouse cages, where Dr. Bengston was directing his attention and healing (**Figure 2**). This turned out to be fortuitous, because we could continue monitoring magnetic field activity next to the mouse cages after he returned to his home in New York, from where he continued sending the mice distant healing.

In later experiments with human subjects (Moga 2014), we found the greatest magnetic signals adjacent to the person being healed, rather than next to the healer (**Figure 3**).

There have been studies demonstrating the ability of healers to alter the output of scientific instruments, but we feel the "electromagnetic signatures" detected in healing sessions are more related to the healing process than to healer ability, which could explain the low number of healers in the Seto study who were able to emit magnetic field waves.

What is the significance of the magnetic field oscillations that we detected in our healing experiments with both mice and human subjects? These signatures may have real importance, as peer-reviewed studies have shown that weak magnetic fields can slow cancer growth and improve survival. In particular, several studies (Buckner et al. 2015; Hu et al. 2010) in the laboratory of Dr. Michael Persinger exposed mice with tumors to a weak, artificial magnetic field called the *Thomas Field,* and they were able to obtain significant tumor regression (**Figure 4**). The Thomas Field is similar to the magnetic field wave we detected when Dr. Bengston treated mice with tumors.

Because healing can occur both locally and at a distance, an effective theory of energy healing must account for nonlocality (action at a distance).

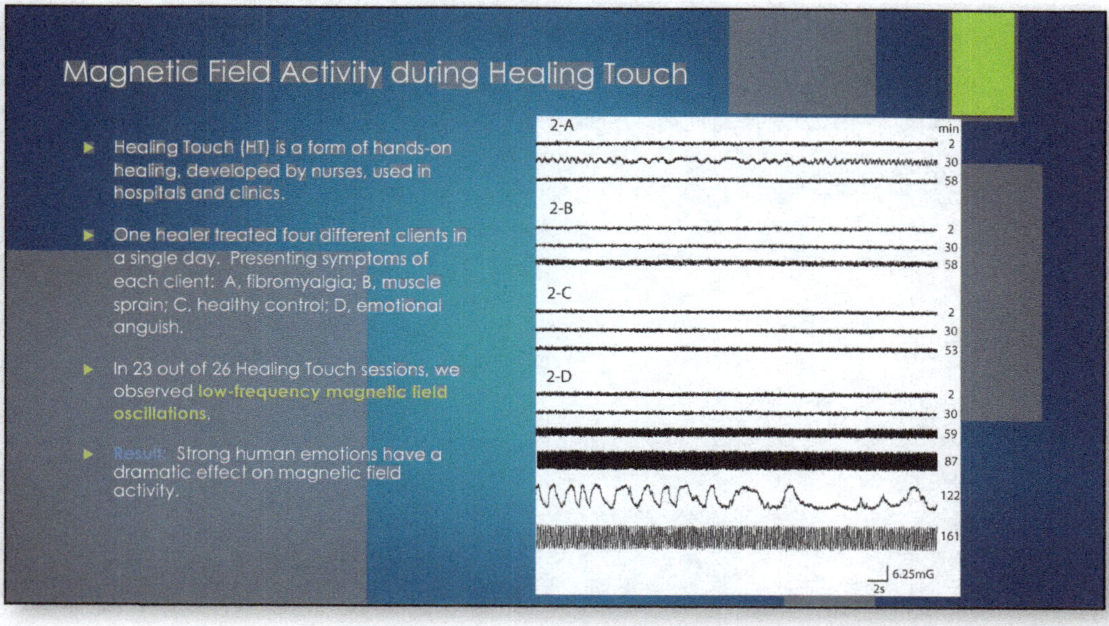

Figure 3

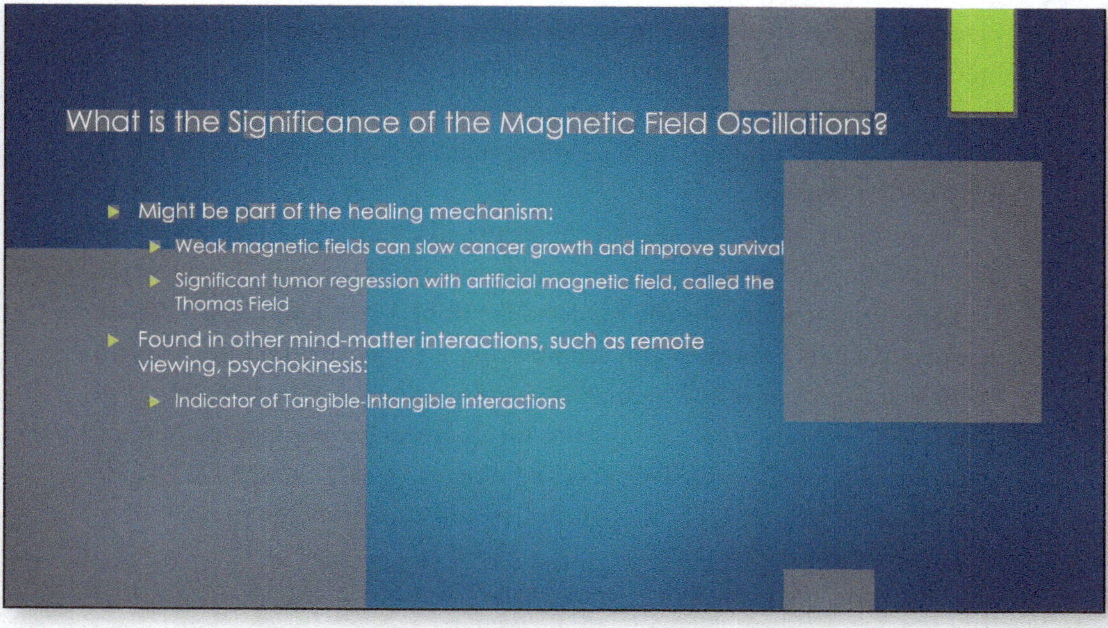

Figure 4

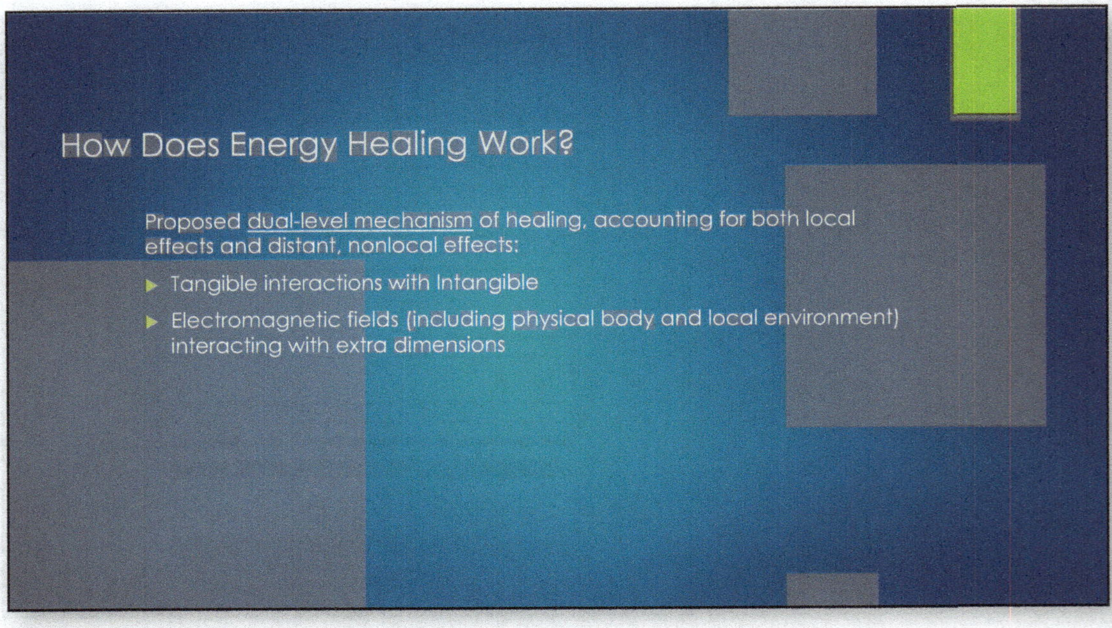

Figure 5

Electromagnetic fields are usually dismissed as part of the energy-healing mechanism because of the inverse-square law.

However, there may be a dual-level mechanism in energy healing: a local, possibly electromagnetic field effect and a distant nonlocal effect, working together (**Figure 5**).

A number of investigators have proposed dual-level mechanisms for mind-matter interactions, such as healing and psychokinesis. For example, Dr. Bengston mentions the M5 model of Dr. Robert Jahn and Brenda Dunne, where conscious processes and tangible space (local effects) interact with unconscious processes and intangible space (nonlocal connections). Dr. William Tiller has outlined a dual-level mechanism for anomalous healing involving reciprocal space (R-Space) and direct space (D-Space). Dr. Claude Swanson has proposed parallel universes and torsion fields to span both local and nonlocal effects.

Based on my experiences with healers and healing, I propose a dual-level magnetic induction/extra-dimensions model of energy healing. The changing magnetic field environment in a healing session can potentially lead to current changes in the client's body (that is, magnetic induction), with beneficial cellular effects (Slater 2009). Regarding extra dimensions, many healers have a sense that the information they receive and work with during a healing session (for example, images, lights, and where to put their hands) exists in a "higher dimension," and that the spiritual Guides, deceased loved ones, and subtle anatomy of the body all reside within higher dimensions. The ability of humans to access higher dimensions may be based on neuroanatomy. The Blue Brain Project (Reimann et al. 2017) recently found that activated neural connections form geometric structures spanning higher dimensions, typically up to six or seven dimensions, with some reaching the eleventh dimension. ◇

References

Buckner, Carly A., Alison L. Buckner, Stan A. Koren, Michael A. Persinger, and Robert M. Lafrenie. 2015. "Inhibition of Cancer Cell Growth by Exposure to a Specific Time-Varying Electromagnetic Field Involves T-Type Calcium Channels." *PLoS ONE* 10 (4). doi: 10.1371/journal.pone.0124136. Accessed on October 25, 2021.

Green, Elmer E., Peter A. Parks, Paul M. Guyer, Steven L. Fahrion, and Lolafaye Coyne. 1991. "Anomalous Electrostatic Phenomena in Exceptional Subjects." *Subtle Energies* 2 (3): 69–94.

Hu, Jing H., Linda S. St-Pierre, Carly A. Buckner, Robert M. Lafrenie, and Michael A. Persinger. 2010. "Growth of Injected Melanoma Cells Is Suppressed by Whole Body Exposure to Specific Spatial-Temporal Configurations of Weak Intensity Magnetic Fields." *International Journal of Radiation Biology* 86 (2): 79–88.

Moga, Margaret M. 2014. "Magnetic Field Activity during Psychic Healing: A Preliminary Study with Healing Touch Practitioners." *Journal of Nonlocality* 3 (1). https://journals.sfu.ca/jnonlocality/index.php/jnonlocality/article/view/50. Accessed on October 25, 2021.

———, and William F. Bengston. 2010. "Anomalous Magnetic Field Activity during a Bioenergy Healing Experiment." *Journal of Scientific Exploration* 24 (3): 397–410.

Reimann, Michael W., Max Nolte, Martina Scolamiero, Katharine Turner, Rodrigo Perin, Giuseppe Chindemi, Pawel Dlotko, Ran Levi, Kathryn Hess, and Henry Markram. 2017. "Cliques of Neurons Bound into Cavities Provide a Missing Link between Structure and Function." *Frontiers in Computational Neuroscience* 11. doi: 10.3389/fncom.2017.00048. Accessed on October 25, 2021.

Seto, Akira, Chikaaki Kusaka, Seiki Nakazato, Wei rong Huang, Takao Sato, Tadashi Hisamitsu, and Chifuyu Takeshige. 1992. "Detection of Extraordinary Large Bio-magnetic Field Strength from Human Hand during External Qi Emission." *Acupuncture & Electro-Therapeutics Research* 17 (2): 75–94.

Slater, Victoria E. 2009. "Energy Healing." In *Holistic Nursing: A Handbook for Practice,* 5th edition, edited by Barbara M. Dossey and Lynn Keegan, 647–673. Sudbury, MA: Jones and Bartlett Publishers.

Session 1 | Second Commentary
On the Presentation by Professor William Bengston

The Flow of Science from Tangible to Intangible Dimensions

Professor GunWoong Bahng
Department of Mechanical Engineering,
The State University of New York, Republic of Korea

I am very pleased to have this opportunity to comment on Prof. William Bengston's presentation on his long-standing research on paranormal phenomena, especially healing.

Dr. Bernard Grad of McGill University in Canada, who died in 2010, conducted a lot of research on remote as well as hands-on healing, with the cooperation of skilled healers in the 1960s. However, his findings were poorly evaluated at that time because of academic prejudice. As Prof. Bengston mentioned, the situation has changed a lot now, as the understanding of paranormal phenomena in the academic world and among the general

About the Speaker

- Leading Professor, Department of Mechanical Engineering, The State University of New York, Korea (SUNY Korea) (2014 - present)
- Principal research scientist, KRISS (Korea Research Institute of Standards and Science) (1989–2013)
- Director, National Center for Standard Reference Data, KRISS (2009–2011)
- Director, Office of Global Partnership, KRISS (2006–2008)
- Author of two books (*New Science Changes the World, Ki Moves the World*), coauthor of eight books, translator of four books – Currently studying water structure and searching for new materials that enhance life energy in people
- PhD in Materials Science and Engineering, Northwestern University

public has been expanding, and academic research results are being reported one after another in academic journals.

Nevertheless, we still do not understand the mechanism of healing so well. Dr. Bengston has worked in this field for a long time, and he summarized the results and presented them today. I would like to comment on two areas of his findings. One is about the perspective in which healing energy is considered as information. The other is about a source of healing energy.

Dr. John Wheeler, who contributed to widespread use of the term "black hole," said in an interview that after he had just completed his degree, he thought the world was made up of particles. Later, his perspective changed from particles to fields, and eventually he reached the conclusion that the world is composed of information. This is a great insight, which means that this universe could be quite different from what we experience with our physical senses.

Nowadays, the wall between East and West is falling, and the old traditions of the East may help broaden our understanding of the universe from the perspective of information. The concept of quantity underlies the concept of energy derived from the West. The focus of energy is on its strength and weakness, as well as more or less. In the East, there was a concept similar to energy for a long time. In India, it was called *Prana,* and in Korea, China, and Japan, it was called *Chi* or *Qi.* Interestingly, the concept of *Qi* contains more information than the strength of energy—it also contains the concept of quality. An example is the division of energy into five "elements," called *O-haeng* in Korean and *Wuxing* in Chinese. According to this theory, energy is divided into the elements of wood, fire, earth, metal, and water. They interact with one another, leading to inter-promoting or inter-regulating effects. If information is combined with the concept of the five elements, a broader understanding may be possible. It means that the quantity and quality of energy can be combined to describe the characteristics of healing energy. For example, healing energy should be inter-promoting rather than inter-regulating energy to promote the recovery of a sick body.

Next, I have some comments on the sources of healing energy. It has been found that healing energy has a feature that transcends time and space.

If so, it should be regarded as a phenomenon outside of the three-dimensional world that we experience. In general, time and space are understood as different from each other, but it is not possible to distinguish between them. For example, when time stops, everything stops. There will be no more changes. What if space freezes? The property of space is distance, so when space is frozen, distances do not change. In this case, everything stops, just as time had stopped. No matter how much time passes, no change can happen. Thus, we may conclude that the reality of this world is change, an illusion, and space and time must change together. Space and time cannot be separated. They are not different, but one.

To make things change, that is, to change space and time together, we need a flow of energy. This flow cannot be supplied from the three-dimensional world itself in which we live. Just as a waterwheel needs to have a flow of water to spin, there must be a flow of energy from a higher level to a lower level in the three-dimensional world. From this point of view and the characteristics of healing energy, transcending space and time, the source of healing energy can be in a higher dimension. The "M5" model of Robert Jahn and Brenda Dunne, a "modular model of mind/matter manifestation," may be interpreted in this way. The importance of *need,* and *intention,* in healing can be found even in the Bible (Matthew 7:7): "Ask, and it will be given to you; seek, and you will find; knock, and the door will be opened to you."

Someone once asked Dr. Grad, "What would you like to say to the world, that you have not yet said?" His reply, "Two and two does not [always] equal four." (Wiand 2010).

As Dr. Bengston said, the equivalence of matter and energy was confirmed in the twentieth century, and now we are entering an era in which the equivalence of energy and information is being confirmed. In the future, I hope that research in this field will continue to deepen our understanding of the universe. ◇

Reference

Wiand, Lenore. 2010. "Bernard Grad: The Legacy Lives On." *Subtle Energies & Energy Medicine* 21 (2): 9.

Session 1 | Presentation Two

Revising Our Concepts About Reality: The Challenge of Consciousness

Dr. Dean Radin
Chief Scientist, Institute of Noetic Sciences,
Petaluma, California, USA

Scientific concepts about the nature of the universe radically evolved over the course of the twentieth century. Advancements in many disciplines, from physics to psychology, caused major revisions in how most educated peoples thought about themselves, their place in the universe, and their understanding of the physical "fabric of reality."

Science will continue to evolve in the twenty-first century, with some of the most revolutionary advancements stimulated by the study of consciousness. Today's key questions about consciousness include: 1) How can the physical brain account for nonphysical qualia (subjective experience)?

About the Speaker

- Chief Scientist, Institute of Noetic Sciences, Petaluma, California
- Associated Distinguished Professor of Integral and Transpersonal Psychology, California Institute of Integral Studies, San Francisco, California
- Founding Board Member, Academy for the Advancement of Postmaterialist Sciences
- Author or coauthor of hundreds of scientific, technical, and popular articles, four dozen book chapters, and four bestselling popular books: *The Conscious Universe; Entangled Minds; Supernormal;* and *Real Magic*
- Editor of the Elsevier journal *Explore* since 2009
- PhD in Psychology, University of Illinois Urbana-Champaign

and 2) How does evidence for exceptional cognitive skills and nonlocal forms of awareness influence our understanding of the mind-brain relationship? Question 1 has been called the *hard problem* of consciousness. Question 2 is what might be called the *harder problem* of consciousness, because it not only questions the neuroscience dogma that the mind is solely a product of brain activity, but it also challenges the set of assumptions that underlies modern science itself—the doctrine called *materialism*.

After Thomas Kuhn's account of the "structure of scientific revolutions" (Kuhn 1970), it is better appreciated that anomalies encountered within a discipline are strenuously resisted by the status quo, often to the point where evidence supporting the existence of the anomalies is viewed as chronically insufficient—or worse, is labeled *pseudoscience*. Eventually, evidence improves and accumulates until it becomes overwhelming and forces a paradigm shift. As anomalies associated with consciousness are better understood, by the turn of the twenty-second century there is a real possibility that the materialistic foundations of science will experience a metaparadigm shift. Rather than being viewed as fundamental, materialism may be reframed as a special case of a more comprehensive worldview, one that views consciousness as fundamental. The scientific, technological, and sociological reverberations of such an ideological shift could dwarf all previous advancements in human knowledge.

Clouds on the Horizon

In April 1900, the Right Honorable Lord Kelvin gave a lecture to the Royal Society (**Figure 1**). In it, the prominent British physicist confidently stated that physics was so successful that it was essentially complete, except for two "small clouds" (Kelvin 1901). The two clouds would come to be known as the *luminiferous ether* and the *ultraviolet catastrophe*. Both referred to anomalies that did not fit the predominant view of what is now called *classical physics*, and both clouds were assumed to be problems that would eventually be solved by minor tweaks to existing theories.

Eight months later, German physicist Max Planck presented a new idea

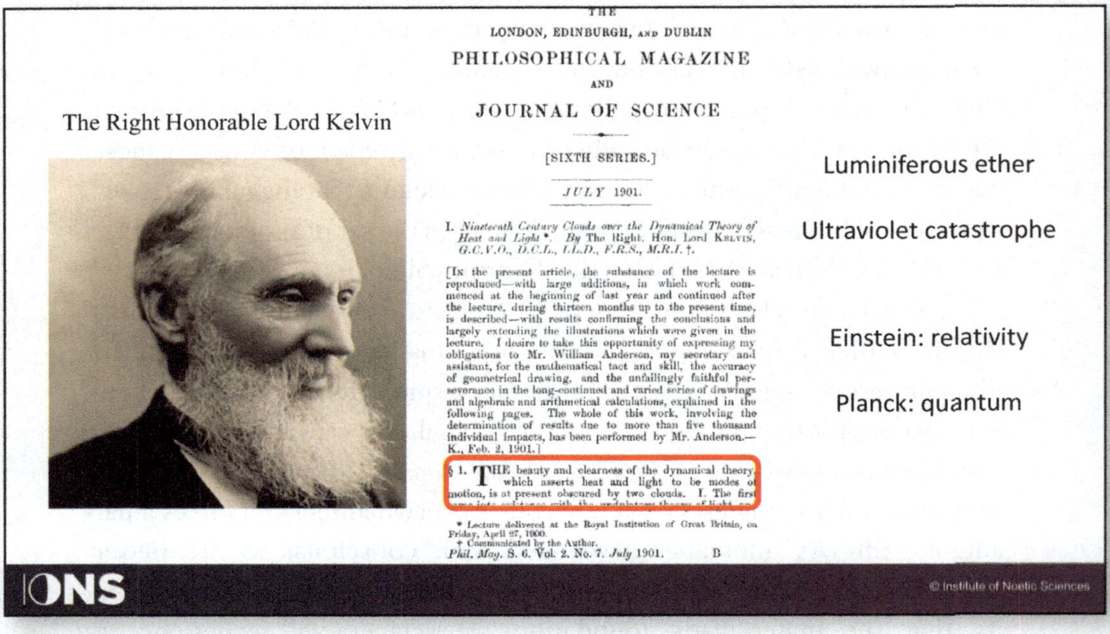

Figure 1

at a meeting of the German Physical Society held in Berlin (Planck 1901). Planck's idea resolved one of the clouds mentioned by Lord Kelvin and founded quantum theory. Albert Einstein explained the other cloud a few years later.

This unexpected turn of events, one of many found in the history of science, serves to remind us why small clouds on the horizon are worthy of very close attention. Sometimes those clouds dissipate after slight revisions to existing ideas. But sometimes they are puzzles that persist for decades or centuries. In such cases, the solutions may stimulate paradigm shifts and usher in startlingly new concepts, technologies, and even new forms of civilization. Over the course of the twentieth century, Lord Kelvin's two clouds transformed the Western world from the industrial age into the atomic and information ages.

Today, we are faced with two other very persistent clouds, commonly known as *qualia* and *quanta* (**Figure 2**). The first cloud has been a mystery throughout history. The second arose as a consequence of Planck's idea. The word *qualia* refers to the nature of subjective experience and *quanta* to the fact

Figure 2

that quantum objects are exquisitely sensitive to being observed. Both clouds raise questions about the nature and role of consciousness in the physical world, and both are major challenges to the scientific paradigm of reductive materialism—the assumption that everything, including mind, consists of matter and energy, and that any system, regardless of how complex, can be completely understood by reducing it to its elementary physical components.

Some neuroscientists insist that qualia constitute a nonproblem because consciousness is an illusory side effect of brain processing (Churchland 1986; Crick 1994). Others propose that any physical system as complex as the brain will spontaneously develop conscious awareness, through some as yet unknown process (Tegmark 2015). Some physicists believe that the quantum observer effect (closely related to the quantum measurement problem) is also a nonproblem, because consciousness plays no role in physics or that the problem is already solved by concepts like decoherence (Schlosshauer 2007).

Like Lord Kelvin, many scientists today undoubtedly assume that these two "consciousness clouds" will eventually be understood in conventional terms. I believe that sentiment is wrong. These two clouds have stubbornly resisted orthodox explanations, and instead of wispy puffs fading away in the light of existing theories, qualia and quanta presage paradigmatic superstorms. They are also the leading edge of a host of related clouds, each more challenging than the last (Schwartz 2010). These include the phenomena of genius, savants, near-death experiences, mediumship, reincarnation cases, and laboratory studies of psychic phenomena. All these phenomena stretch the materialist "brain equals mind" assumption because they suggest that the mind is not inextricably limited to the operations of the physical brain.

Genius

No one who studies the lives and works of Mozart, da Vinci, Copernicus, Shakespeare, Einstein, or Ramanujan can doubt that genius is real, if rare. Of the estimated 100 billion humans who have ever lived, every now and then someone comes along whose talent is so prodigious that it literally shapes the future course of civilization.

The challenge presented by genius is to imagine how the mind, viewed solely as an aspect of brain processing, could generate world-changing mathematical theorems, breakthrough scientific ideas, hypercreative inventions, masterwork books and musical compositions, and the like, apparently arising out of the blue, often unbidden, and fully formed (Schwartz 2010, Heilman 2016). If such ideas arose once in a person's lifetime, perhaps we might dismiss them as a fluke. But true genius is a persistent fount of paradigm-shattering creativity that defies our understanding of mindless electrochemical activity in a brain that is strictly limited to ideas it has already absorbed (Lingg and Frank 1973; Pandey 2001).

Savants

Autistic savants have little to no social skills and low IQs, and yet they can display supernormal capacities of memory, musical talent, artistic talent, or lightning-fast mathematical calculations (Dossey 2012; Cowan and Frith

Figure 3

2009; Welling 1994) (**Figure 3**). The Academy Award-winning 1988 movie *Rain Man* was based partially on the life of savant Kim Peek, who among other things could correctly and instantly recall every word of the estimated 12,000 books he had read. Psychiatrist Darold Treffert, discussing autistic savants, wrote that "Kim Peek possesses one of the most extraordinary memories ever recorded. Until we can explain his abilities, we cannot pretend to understand human cognition" (Treffert 2010).

Treffert also described the case of Leslie Lemke, who "is blind, severely cognitively impaired, and has cerebral palsy. Yet he played back Tchaikovsky's Piano Concerto No. 1 flawlessly after hearing it for the first time at age 14" (Treffert 2010, 288). If one were to test normally healthy pianists who had not previously heard this concerto, it is safe to say that precisely none of them would be able to do this. Treffert goes on to describe the strange phenomenon of *acquired savants,* in which case, as the result of an accident, a normal person suddenly gains savant skills. In addition, there are the completely astonishing cases of *sudden savants,* in which apparently normal people spontaneously gain savant skills, for no known reason.

How the brains of autistic savants work is a major problem for the neurosciences, and perhaps at some point such savants' skills might be explainable via conventional concepts. Yet, how similar skills can arise in acquired or sudden savants remains a baffling mystery.

Psychic Phenomena

There are no broadly accepted explanations for the talents of geniuses and savants, yet they are so rare that they are easy for skeptics to set aside while they instead concentrate on understanding "ordinary" people. This is why commonly reported psychic phenomena, such as telepathy, clairvoyance, precognition, and psychokinesis, are important to consider. These experiences have been reported by ordinary people throughout history, across all cultures, and at all levels of educational experience. Rather than having to rely on astonishing anecdotes for evidence, in this domain a wealth of strictly controlled experimental studies can be found in the peer-reviewed scientific literature (Radin 1997, 2006, 2013, 2018) (**Figure 4**).

Figure 4

Scientific studies of psychic abilities began in the late 1800s. By the 1950s, the accumulated evidence had convinced many academics who had paid attention to the relevant literature. By the second decade of the twenty-first century, the weight of the empirical evidence was overwhelming to all but the most entrenched skeptics. Today, this topic is still regarded as controversial, but not because empirical data are lacking. Rather, the implications of these phenomena are so difficult to accommodate within a materialistic paradigm that critics find it easier to imagine that the evidence must be flawed in some unspecified way, or they insist on a suitable explanation first, before they will even look at the data.

It is beyond the scope of this article to do justice to the wealth of available evidence. However, to illustrate the kind of evidence that is available, we will briefly review one type of telepathy experiment.

Testing for Telepathy

Telepathy involves the experience of gaining an impression about another person's experience, intentions, thoughts, or emotions without the use of the ordinary senses, and without regard to spatial distance or shielding. One of

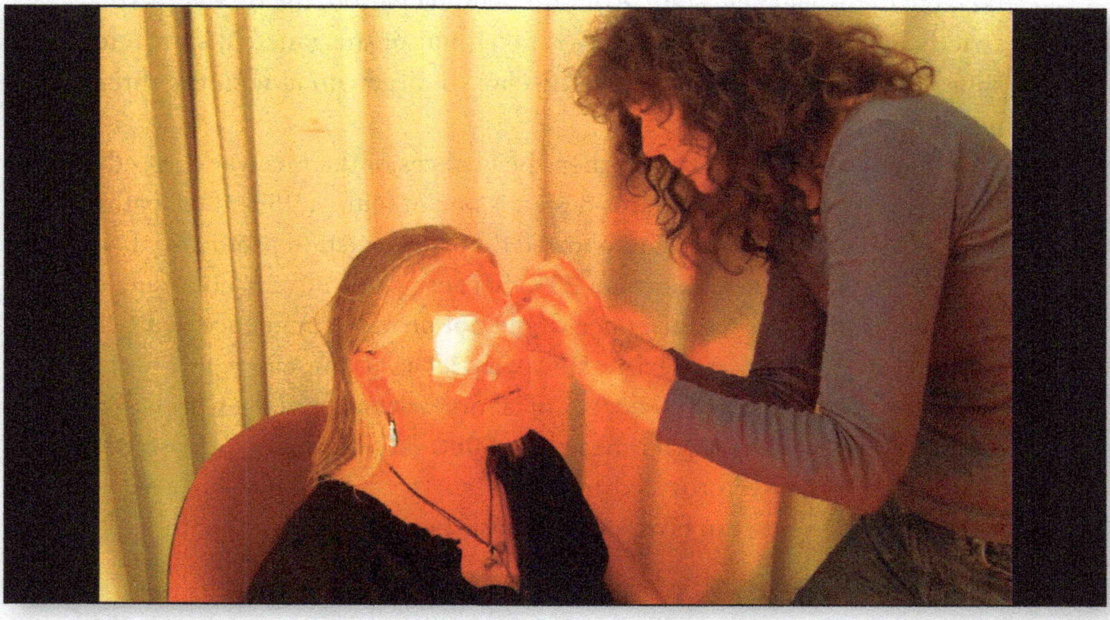

Figure 5

the most successful protocols for testing telepathy in the laboratory is called the *ganzfeld* method (meaning *whole field* in German). In this experiment, a "receiver" of telepathic information has a halved ping-pong ball placed over each eye, the face is illuminated by a soft red light, and white noise is played over headphones (**Figure 5**). This state of mild, unpatterned sensory stimulus induces a dreamlike reverie, which is thought to be conducive to sensing telepathic impressions. While in this state, the receiver is asked to be open to any ideas or feelings gained while holding a distant "sender" in mind.

To conduct the experiment, one photo is randomly selected out of a pool of four photos, where each image depicts a real object or scene with a clearly identifiable theme, and where the colors, shapes, and content of the four photos are as different from one another as possible. The selected photo is given to a sender—who is strictly isolated from the receiver—and he or she is asked to mentally send that photo to the receiver. Note that the use of the words *sender* and *receiver* emphasizes that these terms are descriptive only; they do not suggest underlying mechanisms.

The sender now tries to mentally transmit the contents of the target photo to the receiver for 20 minutes. During that time, the receiver is relaxing in the ganzfeld state. After the sending period, the receiver—still strictly isolated from the sender—is taken out of the ganzfeld state and shown all four photos, one being the chosen target along with the three nonchosen decoys.

If telepathy does not exist, then the chances of the receiver correctly selecting the actual target in this design is one in four, or 25%. If telepathy does exist, and the experiment followed the strict isolation protocols, then the hit rate would be higher than 25%. Because chance is 25% in a single trial, performing this test just once would not provide confidence that telepathy does or does not exist. However, what if the same test were independently performed by dozens of laboratories around the world for a half-century, and during that time nearly 4,000 such tests were performed? The statistical power provided by that many trials would then provide strong evidence either in favor of or against the existence of telepathy.

Meta-analysis

A meta-analysis is a quantitative statistical method for combining the results of numerous experiments based on similar designs. It provides a way to tell if the effects studied in an experiment are repeatable and whether those effects are attributable to chance. Meta-analysis is used in virtually all the experimental sciences today—but especially in the psychological, social, and medical domains because effects in those areas tend to be highly variable, so it is not possible to establish repeatability in a single experiment.

From 1974 to 2018, dozens of authors published 117 articles describing the results of their ganzfeld experiments. Meta-analyses of these studies were conducted seven times, spanning different time scales (Honorton 1985; Bem and Honorton 1994; Milton and Wiseman 1999; Storm and Ertel 2001; Bem, Palmer, and Broughton 2001; Storm, Tressoldi, and Di Risio 2010; Storm and Tressoldi 2020). Each of these seven meta-analyses resulted in independently significant outcomes in favor of telepathy.

This means that repeatable telepathic effects have been observed by dozens of independent investigators around the world for nearly a half-century. Considering all known ganzfeld tests using four targets, some 3,885 tests were reported, resulting in 1,188 hits, for an overall hit rate of 30.6%. Five percent over the chance rate of 25% may not seem very impressive, but from a statistical perspective the overall result is associated with odds against chance at 10,000 trillion to one (**Figure 6, left side**).

One critique often used to discount this result is that some ganzfeld experiments probably failed, which discouraged the investigators from reporting their studies. Such selective reporting would indeed bias the overall result to make it seem stronger than it really was. However, critics who have studied the relevant literature in detail have agreed that selective reporting cannot eliminate the overall positive results. In addition, meta-analytical estimates of the number of presumed unreported "failed" experiments that would be required to nullify the known results confirms that that explanation is implausible (Tressoldi and Storm 2021).

Other critics have questioned whether there might be flaws in the experi-

As of 2020, 117 studies published after the earliest pilot tests

1,188 hits/3,885 sessions
30.6% hit rate

$p = 5.6 \times 10^{-16}$

Odds of 10,000 trillion to 1

Table 6
Artistic participants studies

Study	N	X	X / N	z	$z/\sqrt{N}$
1	20	10	50%	2.20	0.49
2	32	13	41%	1.78	0.31
3	20	6	30%	1.02	0.23
4	97	32	33%	1.67	0.17
5	128	60	47%	5.20	0.46
6	30	14	47%	2.40	0.44
7	40	15	38%	1.64	0.26
Overall	367	152	41.4%	6.83	0.36

$p = 10^{-10}$

Odds of 10 billion to 1

IONS

Figure 6

mental design that would allow the receiver to somehow gain information about the target. Over the years, as critics suggested potential loopholes, each potential flaw was systematically eliminated, and yet the same results continued to be observed. After 50 years of such critiques, skeptics familiar with these studies admit that they can no longer identify any plausible explanations other than telepathy for these results (**Figure 7**). Even skeptics who had explicitly disavowed belief in any sort of psychic phenomena, but nevertheless conducted this experiment themselves, obtained the same results as found in the meta-analyses (Delgado-Romero and Howard 2005).

Conclusion

If the paradigm of reductive materialism does not easily accommodate the challenges presented by the existence of genius, savants, telepathy, and many more examples that could have been mentioned, then what alternative paradigm might be considered? In a nutshell, a viable approach is the philosophical view of *idealism,* which holds that consciousness is fundamental (**Figure 8**).

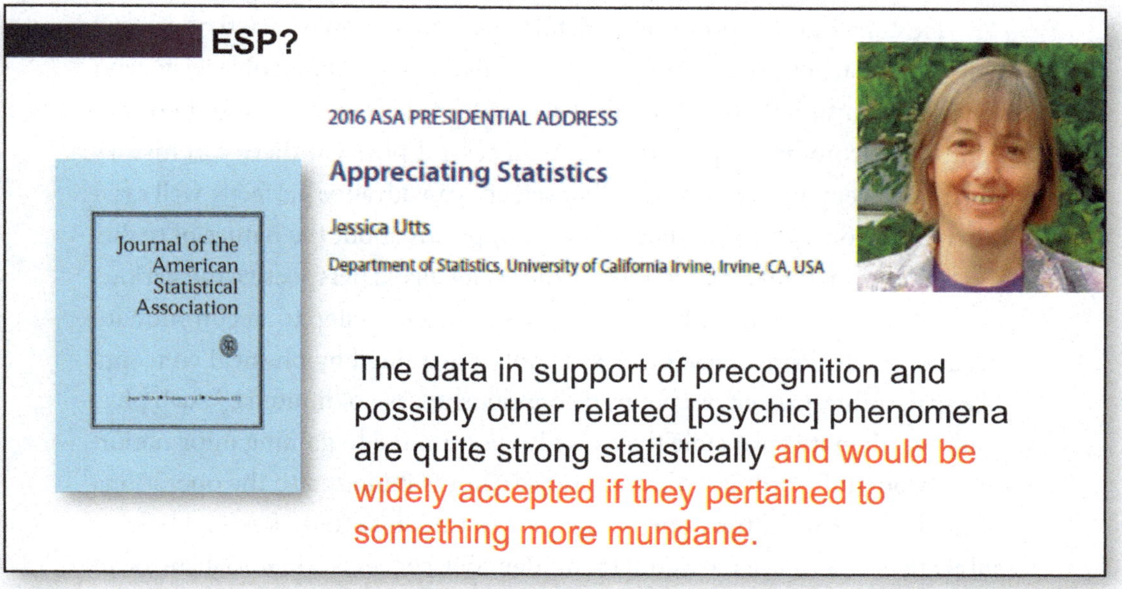

Figure 7

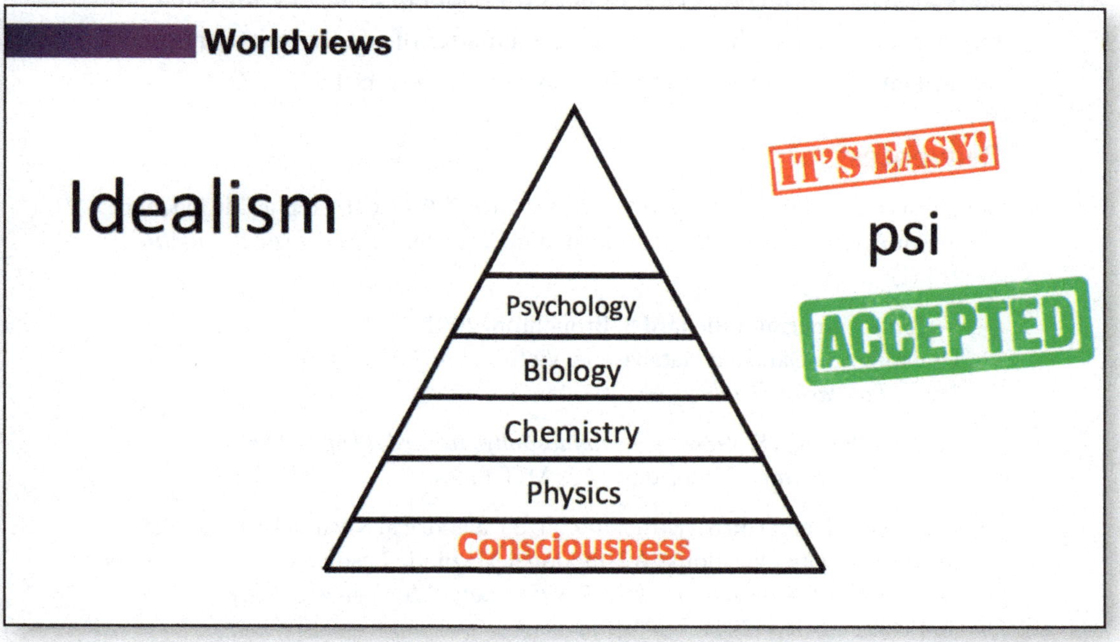

Figure 8

Fleshing out this proposal in detail would take more space than is available for this article, so it simply can be pointed out that most physicists who founded quantum theory were idealists, and yet their worldview did not prevent them from developing the most successful physical theory in history. Their achievements demonstrate that science can advance perfectly well even when based on a very different set of assumptions about the nature of reality.

Moreover, unlike materialism, from an idealistic perspective the various anomalies associated with consciousness are far easier to accommodate, because in idealism, consciousness is not constrained by physical concepts like space, time, matter, or energy. If consciousness is not limited by such physical laws, then it is plausible that it is also not limited to gaining information only through the ordinary physical senses, nor is it limited to the operations of the brain. This in turn opens the door to understanding a whole host of subjective experiences. Despite the undeniable success of materialism as an ideology for understanding the physical world, the empirical and historical facts are that anomalous experiences do happen, even in controlled laboratory experiments. Thus, it is no longer a matter of whether materialism will be supplanted by a more comprehensive worldview, but when. ◊

References

Bem, Daryl J., and Charles Honorton. 1994. "Does Psi Exist? Replicable Evidence for an Anomalous Process of Information Transfer." *Psychological Bulletin* 115 (1): 4–18.

———, John Palmer, and Richard S. Broughton. 2001. "Updating the Ganzfeld Database: A Victim of Its Own Success?" *Journal of Parapsychology* 65: 207–18.

Churchland, Patricia S. 1986. *Neurophilosophy: Toward a Unified Science of the Mind-Brain.* Cambridge, MA: MIT Press.

Cowan, Richard, and Chris Frith. 2009. "Do Calendrical Savants Use Calculation to Answer Date Questions? A Functional Magnetic Resonance Imaging Study." *Philosophical Transactions of the Royal Society B Biological Sciences* 364 (1522): 1417–424.

Crick, Francis. 1994. *The Astonishing Hypothesis: The Scientific Search for the Soul.* New York: Touchstone.

Delgado-Romero, Edward A., and George S. Howard. 2005. "Finding and Correcting Flawed Research Literatures." *The Humanistic Psychologist* 33 (4): 293–303. https://doi.org/10.1207/s15473333thp3304_5.

Dossey, Larry. 2012. "Fractals and the Mind." *Explore (NY)* 8 (5): 263–65. doi: 10.1016/j.explore.2012.06.010.

Heilman, Kenneth M. 2016. "Jews, Creativity and the Genius of Disobedience." *Journal of Religion and Health* 55 (1): 341–49.

Honorton, Charles. 1985. "Meta-analysis of Psi Ganzfeld Research: A Response to Hyman." *Journal of Parapsychology* 49 (1): 51–92.

Kelvin, Right Honorable Lord. 1901. "Nineteenth Century Clouds over the Dynamical Theory of Heat and Light." *The London, Edinburgh, and Dublin Philosophical Magazine and Journal of Science* 2 (7): 1–40. doi:10.1080/14786440109462664.

Kuhn, Thomas. 1970. *The Structure of Scientific Revolutions.* Chicago: University of Chicago Press.

Lingg, Ann M., and Helen Frank. 1973. *Mozart, Genius of Harmony.* Port Washington, NY: Kennikat Press.

Milton, Julie, and Richard Wiseman. 1999. "Does Psi Exist? Lack of Replication of an Anomalous Process of Information Transfer." *Psychological Bulletin* 125 (4): 387–91.

Pandey, Surya Nath. 2001. *Millennium Perspectives on A.K. Ramanujan.* New Delhi: Atlantic Publishers & Distributors.

Planck, Max. 1901. "On the Law of Distribution of Energy in the Normal Spectrum." *Annalen der Physik* 4: 553–58.

Radin, Dean I. 1997. *The Conscious Universe.* San Francisco, CA: HarperOne.

———. 2006. *Entangled Minds.* New York: Simon & Schuster.

———. 2013. *Supernormal.* New York: Random House.

———. 2018. *Real Magic.* New York: Penguin Random House.

Schlosshauer, Maximilian A. 2007. *Decoherence and the Quantum-to-Classical Transition.* Berlin: Springer.

Schwartz, Stephan A. 2010. "Nonlocality and Exceptional Experiences: A Study of Genius, Religious Epiphany, and the Psychic." *Explore* 6 (4): 227–36.

Storm, Lance, and Suitbert Ertel. 2001. "Does Psi Exist? Comments on Milton and Wiseman's (1999) Meta-analysis of Ganzfeld Research." *Psychological Bulletin* 127 (3): 424–33.

Storm, Lance, Patrizio E. Tressoldi, and Lorenzo Di Risio. 2010. "Meta-analysis of Free-Response Studies, 1992–2008: Assessing the Noise Reduction Model in Parapsychology." *Psychological Bulletin* 136 (4): 471–85.

Storm, Lance, and Patrizio E. Tressoldi. 2020. "Meta-analysis of Free-Response Studies, 2009–2018: Assessing the Noise Reduction Model Ten Years On." *Journal of the Society for Psychical Research* 84 (4): 193–219.

Tegmark, Max. 2015. "Consciousness as a State of Matter." arXiv:1401.1219 [quant-ph]. Accessed on November 21, 2021.

Treffert, Darold A. 2011. *Islands of Genius: The Bountiful Mind of the Autistic, Acquired, and Sudden Savant.* Philadelphia: Jessica Kingsley Publishers.

Tressoldi, Patrizio E., Lance Storm, and Dean Radin. 2010. "Extrasensory Perception and Quantum Models of Cognition." *NeuroQuantology* 8 (4, Supplement 1): S81–87.

Tressoldi, Patrizio E., and Lance Storm. 2021. "Stage 2 Registered Report: Anomalous Perception in a Ganzfeld Condition: A Meta-analysis of More Than 40 Years Investigation [version 3; peer review: 1 approved, 2 approved with reservations]." *F1000Research.* https://doi.org/10.12688/f1000research.24868.3.

Welling, Hans. 1994. "Prime Number Identification in Idiots Savants: Can They Calculate Them?" *Journal of Autism and Developmental Disorders* 24 (2): 199–207.

Session 1 | First Commentary
On the Presentation by Dr. Dean Radin

Revising Our Concepts About Reality: The Challenge of Consciousness

Professor Marjorie Woollacott
Institute of Neuroscience,
University of Oregon, Eugene, USA

In his lucid paper about the need to revise our concepts about the nature of consciousness, Dean Radin presents two questions that he calls "persistent clouds" that are major challenges to the scientific paradigm of reductive materialism, regarding the nature and role of consciousness in the physical world. They are: 1) how can the physical brain account for subjective experience, and 2) how does evidence for exceptional cognitive skills and nonlocal forms of awareness influence our understanding of the mind-brain relationship? He notes that these are not small clouds but paradigmatic super-

About the Speaker

- Professor Emeritus, Department of Human Physiology, and a member of the Institute of Neuroscience, University of Oregon, Eugene, Oregon
- President and Founding Board Member, Academy for the Advancement of Postmaterialist Sciences
- Research Director, International Association for Near-Death Studies (IANDS)
- Member of the Steering Committee of the Galileo Commission–Expanding the Scope of Science
- Author of *Infinite Awareness: The Awakening of a Scientific Mind*
- PhD in Cell and Molecular Biology; Research and teaching in neuroscience, child development, aging, rehabilitation, complementary and alternative medicine, and meditation

storms with a host of related clouds, including the phenomena of genius, savants, mediumship, psychic phenomena, and near-death experiences (NDEs). To the list of phenomena that include nonlocal perception (not limited by sensory filters), I would add mystical experiences occurring spontaneously or during meditation.

Radin presents data from research in each of his categories, which challenge the paradigm of reductive materialism. The information builds a strong foundation of evidence establishing the weakness of the reductive materialism paradigm proposing that "brain equals mind." I believe there is an additional level of inquiry that must accompany this foundational information of verified accounts in which individuals are able to perceive information not available through the five senses. This additional level relates to the modulation of hypothesized brain processes through which this might occur.

As early as the late 1800s, William James proposed that the brain acts as a selective filter to the vast amount of information that is theoretically available to be perceived by an individual. As the brain has a limited processing capability, these filters underlie our ability to function successfully throughout our lives. Research by Broadbent (1958) and others (Driver 2001) suggests that filters occur at all levels of the brain, including the sensory receptors themselves, which canalize the vibratory information available to us into perceptual inputs, as well as higher level attentional filters that further limit the information we consciously perceive. These include a wide-ranging default mode network (DMN), also called the narrative or mind-wandering network of the brain. It is the source of the background narratives reverberating continually in our minds, often associated with the egoic notion of self and our roles in the world (Carhart-Harris and Friston 2010).

The DMN is also part of the thalamocortical loop, a reverberating network between higher cognitive centers of the brain and the thalamus, in the midbrain. Research suggests that the thalamocortical loop creates in the brain a potentially "closed" system, with input to the thalamus from the cerebral cortex being larger than that from the sensory systems (Llinás and Paré 1996). This means that normally neural inputs from our current thoughts are more powerful than inputs about our environment coming through our senses in affecting our perception of reality in any moment.

A growing number of neuroscientists (Barrett and Griffiths 2018, Brewer et al. 2011, Woollacott and Shumway-Cook 2020) have proposed that many experiences (such as NDEs and other mystical experiences), which include "nonlocal components" such as out-of-body experiences, and a sense of unity, joy, and expansiveness, are associated with ego-dissolution, that is, with an experience of the border between oneself and the external world dissolving. Supporting this, Judson Brewer has demonstrated significant reductions in two primary nodes of the DMN during meditation, when comparing trained meditators to a control group (Brewer et al. 2011). These results indicate that the brain definitely becomes quieter in meditation, as the narrative part of the brain is silenced.

Does research on NDEs show a similar phenomenon? Brain imaging is not feasible during an NDE related to cardiac arrest. However, available EEG data show that the entire cortex, including the DMN, is deactivated (van Lommel et al. 2001, Woollacott and Shumway-Cook 2020).

What is the significance of this quieting of the DMN? Woollacott and Shumway-Cook (2020) have hypothesized that during NDEs, mediumship, deep meditation, and other experiences involving perception of nonlocal events, a reduction in the activity in the DMN quiets the narrative mind. This results in a diminishing of the normal filtering process of the brain, thus enabling an expanded awareness of subtler perceptions that are normally filtered out. Results from studies of mystical experiences show that the reduction in this activity is directly correlated with an individual's increased sense of unity consciousness and ego-dissolution (Barrett and Griffiths 2018).

These data suggest that the DMN and the thalamocortical loop can be conceived as a reducing valve mechanism that normally confines the contents of a vaster nonlocal awareness within narrow limits defined by the requirements of day-to-day life (Kastrup and Kelly 2018; Swanson 2018). During NDEs and other mystical or anomalous experiences, when the DMN activity is substantially reduced, it is proposed that filtering is lessened, and with this, awareness expands to allow perceptions of these nonlocal phenomena. ◊

References

Barrett, F.S., and R.R. Griffiths. 2018. "Classic Hallucinogens and Mystical Experiences: Phenomenology and Neural Correlates." *Current Topics in Behavioral Neurosciences* 36: 393–430.

Brewer, J.A., P.D. Worhunsky, J.R. Gray, Y.Y. Tang, J. Weber, and H. Kober. 2011. "Meditation Experience Is Associated with Differences in Default Mode Network Activity and Connectivity." *Proceedings of the National Academy of Sciences USA* 108 (50): 20254–59.

Broadbent, D.E. 1958. *Perception and Communication.* Oxford, UK: Pergamon Press.

Carhart-Harris, R.L., and K.J. Friston. 2010. "The Default-Mode, Ego-Functions and Free-Energy: A Neurobiological Account of Freudian Ideas." *Brain* 133 (4): 1265–83. https://doi.org/10.1093/brain/awq010. Accessed on October 4, 2021.

Driver, J. 2001. "A Selective Review of Selective Attention Research from the Past Century." *British Journal of Psychology* 92 (Pt 1): 53–78. https://pubmed.ncbi.nlm.nih.gov/11256770/. Accessed on October 24, 2021.

Kastrup, B., and E.F. Kelly. 2018. "Misreporting and Confirmation Bias in Psychedelic Research: What do images of the brain under psychedelics really tell us about its relation to the mind?" *Scientific American* (blog) Sept. 3. https://blogs.scientificamerican.com/observations/misreporting-and-confirmation-bias-in-psychedelic-research/. Accessed on October 22, 2021.

Llinás, R.R., and D. Paré. 1996. "The Brain as a Closed System Modulated by the Senses." In *The Mind–Brain Continuum: Sensory Processes,* edited by R.R. Llinás and P.S. Churchland, 1–18. Cambridge, MA: MIT Press.

Swanson, L.R. 2018. "Unifying Theories of Psychedelic Drug Effects." *Frontiers in Pharmacology.* https://doi.org/10.3389/fphar.2018.00172. Accessed on October 23, 2021.

van Lommel, P., R. van Wees, V. Meyers, and I. Elfferich. 2001. "Near-Death Experience in Survivors of Cardiac Arrest: A Prospective Study in the Netherlands." *Lancet* 358 (9298): 2039–45.

Woollacott M., and A. Shumway-Cook. 2020. "The Mystical Experience and Its Neural Correlates." *Journal of Near-Death Studies* 38: 3–25.

Session 1 | Second Commentary
On the Presentation by Dr. Dean Radin

Revising Our Concepts About Reality: The Challenge of Consciousness

Dr. Helané Wahbeh
Director of Research, Institute of Noetic Sciences,
Petaluma, California, USA

Dr. Dean Radin, in his presentation titled "Revising Our Concepts About Reality: The Challenge of Consciousness," discusses the challenges arising from the dissonance we find ourselves in today as we straddle two paradigms: materialist and postmaterialist. In one, matter is fundamental; in the other, consciousness is fundamental (**Figure 1**).

If we assume that consciousness is fundamental, then nonlocal consciousness phenomena, where our consciousness transcends conventional notions of time and space, make sense. However, when people have a materialistic and rigid worldview, they cannot integrate new information that contradicts that worldview. They cannot

About the Speaker

- Director of Research, Institute of Noetic Sciences, Petaluma, California
- Adjunct Assistant Professor, Department of Neurology, Oregon Health & Science University, Portland, Oregon
- She has published on and spoken internationally about her studies on complementary and alternative medicine, mind-body medicine, extended human capacities, stress, posttraumatic stress disorder, and their relationships to physiology, health, and healing
- Author of *The Science of Channeling: Why You Should Trust Your Intuition and Embrace the Force That Connects Us All*; coauthor of *Free To Be Me: A Journey of Transformation through Generational Healing*
- Clinically trained as a naturopathic physician (ND) and research trained with a Master of Clinical Research (MCR)

Figure 1

accept evidence outside the materialist paradigm, despite its rigor and repeatability.

The rejection of evidence that supports consciousness as fundamental, or at a minimum that materialism does not explain everything, is not just personal. It extends to active discrimination and suppression of research exploring questions about nonlocal consciousness and its implications. As one author stated, "There is surely no field of study in which the concepts, beliefs, and biases of our 'common-sense' world clash so violently with the data collected and analyzed by scholars as psychical research, or as it is frequently termed today, parapsychology" (Ashby 1987).

Here is one first-hand example of this discrimination. We surveyed meditators about their exceptional experiences during meditation, such as telepathy or non-physical entity communications (**Figure 2**).

We submitted the manuscript for peer review but received a long, passionate letter from the editor describing why the paper was rejected. The letter stated, "Is it worth spending (public) research money and resources, as well as the scientists' time necessary for peer review (let alone misleading

One study surveyed about 1100 meditators from around the world regarding their exceptional experiences during meditation. **Figure 2**

the public's and the media's attention) to test the hypothesis that 1+1 = 3? Or the hypothesis that pigs can fly? Or that water can be turned into wine?... I will do everything in my power to avoid any public research grant money being spent in that direction." Apparently, even asking the question about these experiences during meditation is not scientific because they are impossible.

This sentiment is almost laughable, considering the numerous studies worldwide demonstrating that these experiences are not only possible but incredibly common. Interestingly, the prevalence of exceptional experiences like telepathy was not statistically different in scientists and engineers (59.2%) than the general public (52.7%) (Wahbeh et al. 2018) (**Figure 3**).

Despite intense taboos, scientists report having experiences that are better explained with a postmaterialist paradigm. Unfortunately, the taboos are real, and many academicians do not research these topics or express support for them for fear of being severely criticized or losing their positions

Telepathy in 59% of Scientists

Table 2. Self-report endorsement of EHE by group, sorted by combined prevalence largest to smallest

Question Asked in Survey	General Population	Scientists	Enthusiasts	Pearson $X^2(2)$ or $F(2,897)$, P-value	All Groups
Endorsed ≥ one experience	94.0% [c]	93.2% [a]	99.3% [a,c]	21.1, <0.0005*	96.4%
Mean number of experiences endorsed	7.7 (5.2) [c]	8.7 (5.6) [a]	13.1 (5.8) [a,c]	92.0, <0.0005*	10.5 (6.2)
Total EHE score (0–1.0)	.18 (.14) [c]	.21 (.16) [a]	.32 (.17) [a,c]	83.3, < 0.0005*	.25 (.17)
7. Felt another person's emotions?	65.8% [c]	70.9% [a]	88.6% [a,c]	58.0, <0.0005*	78.0%
6. Just known something to be true or had a clear sensation or feeling of knowing something that you would otherwise have no way of knowing?	69.6% [c]	67.0% [a]	87.4% [a,c]	46.4, <0.0005*	77.8%
16. Known you were dreaming during your dream or been able to control your dreams?	61.5% [c]	63.6% [a]	78.7% [a,c]	29.1, <0.0005*	70.3%
25. Felt as though you were in touch with someone when they were far away from you?[g]	52.7% [c]	59.2% [a]	79.9% [a,c]	64.1, <0.0005*	67.2%

IONS

Figure 3

(Traxler et al., 2012). The article "The Unbearable Fear of Psi: On Scientific Suppression in the Twenty-First Century" (Cardeña 2015) gives many examples of discriminatory treatment of researchers in this field.

Dr. Radin describes some evidence about telepathy, one phenomenon with solid evidence. Remote viewing is another area. Through remote viewing, people can access information from a distant location that they have never been to before and is unavailable to them in traditional ways. Formal remote viewing meta-analyses report some of the most substantial effects in parapsychology with effect sizes ranging from 0.17 to 0.39 (**Figure 4**) (Baptista, Derakhshani, and Tressoldi 2015; Cardeña 2018; Dunne and Jahn 2003; Milton 1997).

Perhaps the most compelling evidence for remote viewing is its practical application. The United States government funded a 20-year secret program centered on remote viewing. Drs. May and Marwaha, who compiled all the Star Gate material after its disclosure (**Figure 5**), commented on the results: "In a total of 504 separate missions from 1973–1995, remote viewing pro-

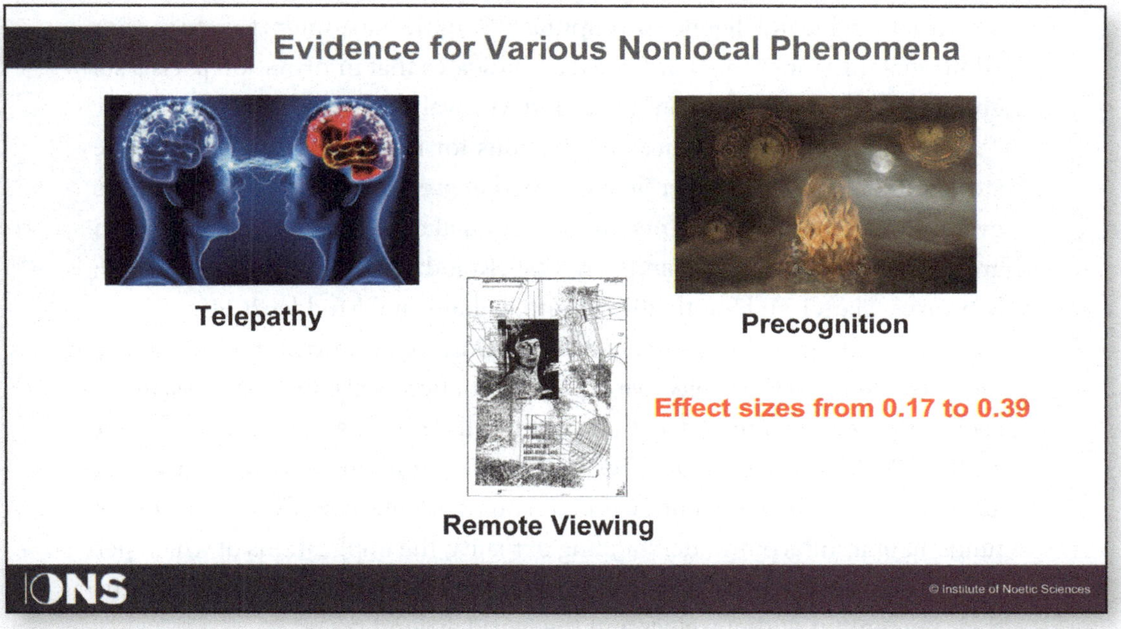

Figure 4

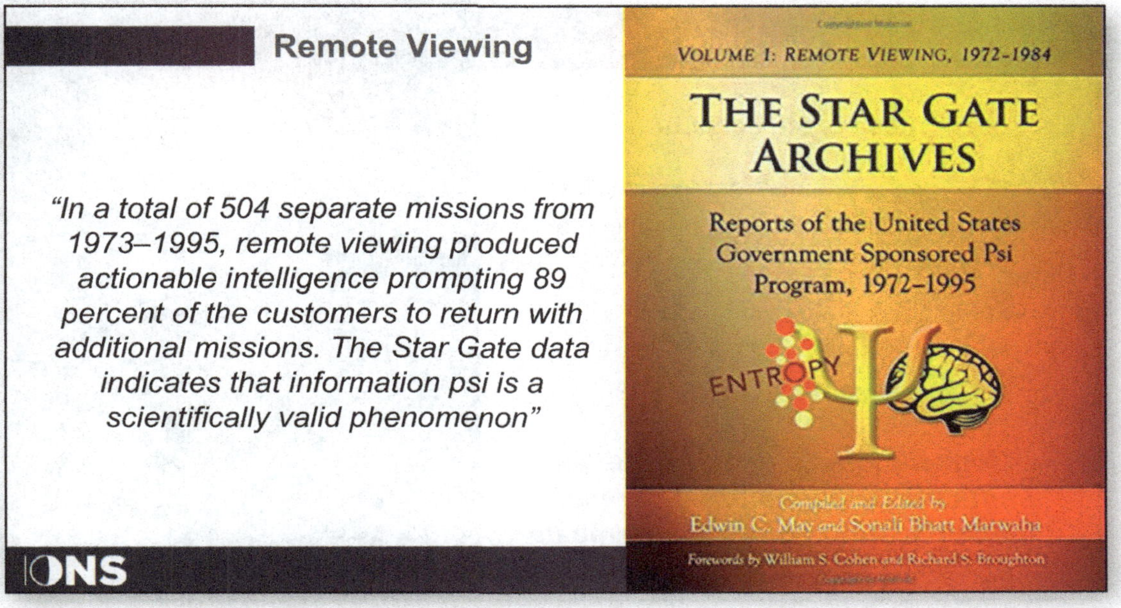

Figure 5

duced actionable intelligence prompting 89% of the customers to return with additional missions. The Star Gate data indicates that information psi is a scientifically valid phenomenon" (May and Marwaha 2018a, 2018b).

Other successful practical applications for remote viewing include the stock market, futures or other financial market predictions, outcomes of sports events, locations of missing persons or criminal cases, and finding unknown archaeological sites (Harary and Targ 1985; Kolodziejzyk 2013; Schwartz 2019; Schwartz, Mattei, and Smith 2019; Smith, Laham, and Moddel 2014).

Even with growing evidence in the laboratory and real world, and the high prevalence of the subjective experience of nonlocal consciousness, extensive resistance remains to what Dr. Radin calls "revising our concepts about reality." Perhaps the difficulty lies in that the paradigm shift to postmaterialism is not just in one scientific domain but in *all* of them. Consciousness as fundamental shifts our understanding of reality, the implications of which permeate every aspect of our lives (**Figure 6**). People resist because it threatens their perception of themselves and the world around them.

Figure 6

Perhaps we can invite resisters to conduct a thought experiment, imagining what the essential research questions would be if consciousness were fundamental. Perhaps holding the hypothetical assumption that consciousness is fundamental and focusing on what we can learn about the mechanism, mediators, moderators, and practical applications of nonlocal consciousness will help us jump across the paradigm chasm we are straddling. ◊

References

Ashby, R.H., and F.C. Tribbe. 1987.
The Ashby Guidebook for Study of the Paranormal.
York Beach, ME: S. Weiser, in collaboration with Spiritual Frontiers Fellowship.

Baptista, J., M. Derakhshani, and P.E. Tressoldi. 2015. "Explicit Anomalous Cognition: A Review of the Best Evidence in Ganzfeld, Forced Choice, Remote Viewing and Dream Studies." In *Parapsychology: A Handbook for the 21st Century,* edited by E. Cardeña, J. Palmer, and D. Marcusson-Clavertz, 192–214. Jefferson, NC: McFarland & Company.

Cardeña, E. 2015. "The Unbearable Fear of Psi: On Scientific Suppression in the 21st Century." *Journal of Scientific Exploration* 29 (4): 601–20.

———. 2018. "The Experimental Evidence for Parapsychological Phenomena: A Review." *American Psychologist* 73 (5): 663–77. https://doi.org/10.1037/amp0000236. Accessed on November 8, 2021.

Dunne, B.J., and R.G. Jahn. 2003. "Information and Uncertainty in Remote Perception Research." *Journal of Scientific Exploration* 17 (2): 207–41.

Harary, K., and R. Targ. 1985. "A New Approach to Forecasting Commodity Futures." *PSI Research* 4 (3–4): 79–88.

Kolodziejzyk, G. 2013. "Greg Kolodziejzyk's 13-year Associative Remote Viewing Experiment Results." *Journal of Parapsychology* 76: 349–68.

May, E.C., and S.B. Marwaha, eds. 2018a. *The Star Gate Archives: Reports of the United States Government Sponsored Psi Program, 1972–1995.* Volume 1: *Remote Viewing, 1972–1984.* Jefferson, NC: McFarland & Company, Inc.

———. 2018b. *The Star Gate Archives: Reports of the United States Government Sponsored Psi Program, 1972–1995.* Volume 2: *Remote Viewing, 1985–1995.* Jefferson, NC: McFarland & Company, Inc.

Milton, J. 1997. "Meta-analysis of Free-response ESP Studies without Altered States of Consciousness." *Journal of Parapsychology* 61: 279–319.

Schwartz, S.A. 2019. "The Location and Reconstruction of a Byzantine Structure in Marea, Egypt, Including a Comparison of Electronic Remote Sensing and Remote Viewing." *Journal of Scientific Exploration* 33 (3): 451–80.

Schwartz, S.A., R.J. De Mattei, and R.C. Smith. 2019. "The Caravel Project." *Zeitschrift Für Anomalistik* Volume 19: 113–39.

Smith, C.C., D. Laham, and J. Moddel. 2014. "Stock Market Prediction Using Associative Remote Viewing by Inexperienced Remote Viewers." *Journal of Scientific Exploration* 28: 7–16.

Traxler, M.J., D.J. Foss, R. Podali, and M. Zirnstein. 2012. "Feeling the Past: The Absence of Experimental Evidence for Anomalous Retroactive Influences on Text Processing." *Memory & Cognition* 40 (8): 1366–72. https://doi.org/10.3758/s13421-012-0232-2. Accessed on November 8, 2021.

Wahbeh, H., D. Radin, J. Mossbridge, C. Vieten, and A. Delorme. 2018. "Exceptional Experiences Reported by Scientists and Engineers." *Explore* 14 (5): 329–41. https://doi.org/10.1016/j.explore.2018.05.002. Accessed on November 8, 2021.

Session 1

Discussion

Cassandra Vieten (Session Chair): I have a couple of questions of my own, and then I would like to open it up for the audience to ask questions. My first question is for Dr. Bengston and Dr. Moga: What do you make of this finding that the electromagnetic oscillations are stronger closer to the healee than to the healer?

William Bengston: I can weigh in with a quick comment. I think that healing is driven by the healee. I do not think that healing has a psychokinetic (PK) equivalent in past psychology. In the PK world, you are producing something or doing something. You are the source of it, and then there is an effect. I think that healing is a healee-based phenomenon. That is where the need comes, and the healer is closer to being a transmitter of information. That is not particularly satisfying, because I do not know how it is transmitted, but I would say that the key is the healee, or the biological need.

Margaret Moga: I would agree with that. Also, the healer's intent is on the healee and that is where the healing, or "nonlocal event," is occurring. The oscillations are a perturbation in space, where change is occurring, and that is going to be close to the healee.

Vieten: Thank you. Okay, I have a question for Dr. Bahng. Dr. Bahng, when you said you think that healing is more of an inter-promoting than an inter-regulating process, what is the difference? What does that mean?

GunWoong Bahng: Okay, the difference is as follows. When we study about the effect of healing energy, most people think that the healing energy is productive or helpful for the healee, or the client. But through our experience from a wide area of the effect of the parapsychological field, we find that there is also another side effect, such as suppression while keeping some bad intention to some people, or manipulation, or something like that. So that is what I wanted to say.

Vieten: Thank you. I appreciate that. I have a question for Dean from Marjorie. In your mind, what happens when a person becomes a savant after an NDE, an illness, or an accident? Do you think the filters are reduced momentarily, and the experience may give them a pathway to reduce the filters in future moments? Does it reorganize their brain? Does it open up a chink in the matrix? What do you think?

Dean Radin: I have no idea. We ask neuroscientists how this would be possible, or in worse cases such as terminal lucidity, how this would be possible. We have no idea.

Vieten: Okay, thank you. Very unsatisfying answer, Dr. Radin. It might also be a challenging answer. Dr. Bengston, if it has to do with biological need, why is it that some, let us say a child or a baby of a mother who needs healing tremendously, and the mother would be predetermined to have a strong biological need to heal the baby, why does it not work for all healees?

Bengston: You are presupposing that it does not work, and we do not know that for sure. I will go with Dean, too: "I do not know" is going to be the answer to almost any question you can throw my way. For example, I cannot account for the internal, intragroup variance. If I have, for example, two mice in the same cage with the same healer, they are essentially clones of each other—everything is identical. One mouse will take x number of days to be cured, and the other mouse will take 2.5x. That is huge! You call it unexplained variance, and then you are supposed to breathe a sigh of relief. But the bottom line is, I do not know. Why the range of responses? I do not know.

Moga: I have a thought on this. I am thinking of Dr. Brian Weiss, who lost his son in early childhood. It was not about the child's need to be healed, it was about his need. Weiss needed to learn the postmaterialist view of life, which led to his new life path. So, it was his need for healing that was connected to the unfortunate event.

Vieten: Interesting, thank you. We have a few people asking to speak. How about Richard Lewis?

Richard Lewis: This is about your concept of consciousness being on the level below physics, where it is put at the bottom of the pyramid. The question is, does that lead to the idea of a creative God from whom the universe sprang? Is that a logical connection, or is that not a connection that can be made?

Vieten: Could you restate the first part of your question? I don't think it was audible.

Lewis: Quite a few speakers talked about the pyramid of physics, chemistry, and biology, where consciousness was placed at the bottom-most level. I wonder if that led to the idea of a creative God, a creative Being out of which the universe sprang. I mean, does it lead to this concept?

Bengston: I can say that I stole that pyramid from Dean, and then he used it in his talk. So, I am going to yield to his analysis of this.

Radin: What I would say is that in some traditions, in esoteric traditions, they do talk about a Universal, or a God. But it is not true in all esoteric traditions. So, people can interpret it in different ways. I interpret it to mean that consciousness is part of the fabric of reality. To go beyond that into something like a creative force or entity, or to personify that consciousness? I do not need to go there personally, but I can understand how some might.

Vieten: Thank you. How about William Keepin?

William Keepin: My question is for Dr. Bengston. How effective have your healing methods proven to be for healing cancer in humans?

Bengston: Clinical studies are far more difficult. I began when I started this "100 years" ago, doing humans. That led me to extreme frustration because I could not control all the variables. If a person comes in and you do whatever you do, however many times you do it, and they leave and get better, there are too many variables. You do not know if they changed their diet, they changed their exercise, and so on. I needed to be more of a control freak. But I can tell you anecdotally, many, many humans have claimed to have been fixed of cancer. I also have no way to gauge the success rate, because, when you have humans, you have what is known as a "file drawer problem." What happens to the people who come and then disappear? The answer is, I do not know. In the case of mice, I know where they are, I know their entire lifespan, I know everything about them. That kind of thing is appealing to me. So, there is no way for me to give even a speculative percentage on what the success rate would be. It seems loosely to follow the patterns of mice, but humans, obviously, are a tad more complicated. But in terms of some of the patterns, like more aggressive cancers heal faster, that seems to hold up for humans. But there is no way to put a mathematical equivalent to it.

Vieten: Yes, and I think that in the early days of some of the distant healing research, there was so much excitement about initial promise that a race to large-scale clinical trials took place, many of which ended up negative, and it kind of torpedoed the field. So, it is important to follow the stages of basic research to controlled laboratory research, before you move into the very wild mess of human subjects' trials. Okay, Takahiro Hiroi.

Takahiro Hiroi: I have a question similar to that of Dr. Richard Lewis about Dr. Radin's diagram, showing that consciousness is more fundamental than physics. Actually, physics has gone into very spiritual areas these days. Beyond our known four-dimensional spacetime, they claim based on superstring theory, we need six additional hidden dimensions. So, if those hidden dimensions could carry our consciousness power, like spiritual power or beyond, I suppose that could explain the interactions of human consciousness and between human consciousness and material bodies, as in healing. Do you have any comment on such ideas?

Radin: Given that so far there is no way to definitively test whether string theory is correct, or even if there are multiple dimensions, it sits within the realm of theoretical possibility, but I do not think we can go very much beyond that. I tend to be an empiricist. If somebody presents an interesting idea, if I cannot figure out a way of testing it, it goes on the shelf as just one more theory. There are plenty of theories. To suggest that the twelfth dimension is consciousness, that does not really mean anything to me, because it does not say anything about what consciousness actually is. Once we figure out how to test these ideas, I will be a bit more comfortable saying something about it.

Bengston: I would agree with that, and also point out the importance of predictability. If you simply have a speculative idea, it may sound interesting and all that, but does it lend itself to predictive hypotheses?

Vieten: Thank you. We have a hand raised by another pioneer in the field, Beverly Rubik.

Beverly Rubik: I am interested in signals associated with healing research. As you may know, Dr. William Tiller and Dr. Joie Jones, among others, have said, when we do healing experiments over and over again in our laboratory, things change—a space becomes conditioned, and it gets easier and easier to get positive results in our experiments. So, I am quite intrigued that there were the oscillations in the ambient magnetic field and healing studies with Bill Bengston and also Healing Touch. I wonder if you looked long term, if you have done a sufficient number of experiments, does that oscillating ambient magnetic field remain in the room after the healing experiment, and is it ongoing after a lot of experiments? I am looking for a signature for conditioned space, because as far as I know, we do not have one.

Bengston: I can weigh in on that briefly. Beverly, it is a great question, and it is something that deserves a whole lot more interest. In Providence, Rhode Island, a couple years ago, Dean and I were working together on a project of recording healing intention and playing it back to cells in mice and such, to see if we could reproduce the effect. One of the things in one of the papers we published on the effect of the recording of healing intention on cancer cells was as follows. We examined 167 genes and found a reliable effect of changes in 68 of them. One of the interesting phenomena was related to some "sensitives," that is, people who allege to be sensitive and aware of things that an ordinary person would not be. When they hear the recording and you turn it off, they say it is not off. I keep getting that response: "It is not off." I would hold up the plug and say, "Come on, I have unplugged the machine, you know, it is not playing." They go, "No." They do not even back down. They say, "No, it is still playing." And I go, "You know, this is crazy talk because the machine is not on."

So, I phoned up to the lab and I said, we have got to test this, we have to find out who is crazy here. It turned out I was the one who was crazy. We played the recording to an incubator, then waited, turned it off, unplugged it, and then waited again. Three days later, we stuck cancer cells *in vitro* into the incubator, and they still changed genomically. Now, one of the questions is, how long would that effect last, and we have not done that. But we had clear indications of genomic changes—the space had been conditioned.

This leads to some really interesting issues—if an experimental space can be conditioned, and you have something along the lines of a lag time, or the thing remains. Imagine how many misinterpretations of data are going on, where you think "this" is going on, but actually "that" is going on? In the same way, we have assumptions of locality. If I take subjects—such as mice or people—and I put them in various places, physical separateness is thought to mean independence. Well, what if that is wrong? What if physical separateness does not mean independence? What if you turned something "off"—you have a device going, what if it remains "playing" at least in the room, think of all the data misinterpretations you would have.

Vieten: Dr. Wahbeh, I am curious if you have any thoughts about conditioned space?

Helané Wahbeh: Yes, I was just going to add something. We just finished a large energy medicine study with 17 different practitioners and about 200 people who had hand and wrist pain. We included in our study design a clairvoyant seer who observed all the sessions, and one of her important roles was to clear the space between each session. Now, we did not exactly know how to measure if the space was cleared. However, from her energetic sensitivity, she would say, "Okay, now it is cleared, and let us move on." So, back to Beverly's question, it would be wonderful to discover a device, a measurement that we could take to see if this space is conditioned or not and if it is cleared. It sounds like Dean has something he wants to add about that. Thanks.

Radin: One of the things we did in the experiment that Helané is talking about is, we ran a system that had 16 random number generators running in parallel. We did find that during the healing sessions, there was a significant deviation, meaning that there was a "neg-entropic" effect picked up in the environment. But as a control, we ran the same data eight hours later, in the same room, and we found nothing. So, it kind of suggests that if healing is changing the environment, it can be unchanged—it can be put back to normal, and that is what the seer was doing. So, we do not know if it would have persisted over time, because it kept getting reset each time.

Wahbeh: That was with random number generators. There are other measures that were mentioned in these different talks, and who knows if we had other objective measures what they would show us. So, it is a fascinating, open field.

Vieten: Thank you. Great question, Beverly. How about Roumiana Tsenkova?

Roumiana Tsenkova: Absolutely interesting, fascinating session. I have received some answers to some of my questions. But now, regarding the discussion about space, I want to share what we have been doing in my lab analyzing spectra of water under perturbations. We can monitor the water spectrum in real time, and water is very sensitive to electromagnetic fields. Also, it responds to temperature, humidity, and even subtle influences. It shows a specific spectral pattern for each perturbation. So, if you have spectral monitoring of water in the space (or room), that might be a very good source of information, and it might help with understanding the discussed phenomena. We have monitored the spectral changes of water in living systems, including humans under healing conditions. We could see differences, and a very good correlation between healees' and healers' spectral changes over the time of interaction. So, it might be a very good idea to apply Aquaphotomics (all about water–light interactions) in this field.

Secondly, we heard about consciousness as being fundamental. We are very close to the right answer, but I think we overuse this word. It seems that when using the word "consciousness," we want to cover everything beyond the human interface. Everything that we cannot explain, we think consciousness can. It is much more than that. So, thank you very much again. Also, I thought about perturbations and telepathy. Does telepathy always need external perturbation, or can we register it without perturbation? That is a question I still have. Thank you.

Vieten: Thank you. Yes, I know that the "extraordinary healers" study at the Institute of Noetic Sciences did investigate water, and I believe that they recently may have published their results. Would Dean like to weigh in on that briefly?

Radin: We had two different tests with the healers. We had the healers and the clients (or participants), all of whom wore a necklace with a small aliquot of distilled water, and we measured the spectrum of the water before and after the healing session. We found a significant change in the spectrum of the aliquot that was around the healer's neck but found no change in the water around the participant's neck, which was interesting because they were only about two feet apart. The other thing we did is that we set up a sample of water and had the healer directly heal the water, and there we did see a significant effect. We only saw this in the distilled water and not in commercial spring water. So, it is true that we did find changes in the water. It is not really that convenient, because you have to run the spectrum. So, we cannot do it in real time. Nevertheless, we did see a difference in the water.

Vieten: Yes, and I do appreciate the comment about making sure that we stay somewhat precise about the use of the word "consciousness"—that it does not just become a sort of bucket for the Alpha, the Omega, the everything we cannot explain. It is a tricky thing. I think some of the themes that ran through the papers in the presentations were the ideas of qualia and quanta. Maybe qualia is what we are starting to call consciousness, as opposed to the hundreds of definitions of consciousness—being awake and things like awareness. I thought there was a really interesting tension between Bill Bengston's presentation saying that awareness, at least conscious awareness, is not necessary for any of this, whereas a lot of people would say that consciousness is what it is all about, or the direction of consciousness. So, Dr. Woollacott, would you like to weigh in on anything that has been said so far?

Marjorie Woollacott: I would like to weigh in on what Dean was just talking about—in your experiments, Dean, with the water and the change in the absorbance at various frequencies. I was thinking about all the research on homeopathy and the controversies about whether homeopathy works or does not work. I am wondering if you might give us your opinion about whether perhaps something is going on with the homeopathic making of these liquids, which have no molecules left of the original substance, that allows the liquid to nevertheless have a healing effect. Could the effect be due to similar energetic transfer effects as those in the experiments you discussed?

Radin: Yes, it may well be. It is also true that we use the word "consciousness" as kind of a magic sauce. These kinds of phenomena do not make much sense from a materialistic point of view. But with one exception: a lot of the strangeness of quantum mechanics is similar in kind to what we see here. It may turn out that, as we look back in history, our notion of what we mean by "material" has significantly changed over time. Quantum mechanics is not the end of physics—it will eventually be supplanted by something else. That something else may be closer to what we think of as subjective awareness. Or, if we can start seeing that, as we already see in some biological systems, that there are quantum effects within those systems. I think it is quite likely that there are aspects of the brain that are working at the quantum scale. If that is the case, it means that we as creatures have evolved over millions of years, with quantum phenomena in us that we find very difficult to measure at this point. Except that, because the brain may be giving rise to something like awareness through lots of feedback loops in the structure of the brain, some elements of the brain would be nonlocal. In a quantum sense, it is not as though we are magically anointed with quantum mechanics—we would have been born with it, we evolved with it. So, some elements of what we think of as our own awareness would have quantum properties. You can start listing all the various kinds of psychic or mystical experiences that begin to make sense from a "neo-materialistic" perspective, or a more sophisticated way of thinking about materialism. So, we do not need post-materialism. Or we can say that post-materialism is super-quantum mechanics, or sub-quantum mechanics, or whatever the next stage in physics is.

Woollacott: I could maybe add one question on top of that related to this nonlocal phenomenon, and quantum mechanics may be a part of a lot of people's experience. I was thinking of your comments about genius and about Mozart. When we hear about Mozart, we hear that he was given, in effect, symphonies completely whole, all at once, and it seems to be something of a nonlocal phenomenon, beyond time and space. I just wanted to ask if you have any comments on that.

Radin: Well, genius is difficult, right? It is true that many of the great composers would write down what they internally heard and then not change anything, which is crazy. Nevertheless, maybe they are getting it from their own future. Maybe they are getting it out of some space in the Akashic record. They are able to access areas of space and time that are simply not right here, but they can get it. So, their special genius is, unlike people who are not composers, they are able to get that information and cast it into terms that turn into music, whereas a mathematician might pull something down from who knows where and it turns into a new theorem. So, you need a certain kind of special expertise to be able to interpret that information. I kind of think that most people can get these downloads, but most people do not know what to do with them. They may say, "What a funny dream I had!" But a mathematician would have the dream and say, "Oh, that is a solution to this problem."

Vieten: Yes, and it may be there is some possibility that everyone has access, but we have evolved to have very strong filters, so that we can make it through every day without going crazy. So, it is not so much about the ability to access as the ability to remove the barriers to access, or to inhibit the barriers.

Radin: And without going crazy.

Bengston: You need a social context. So, Dean comes up with this brilliant, innovative idea, and nobody can recognize it. Well, he is not a genius, he is quirky. This definition of genius is a social label. It means that there has to be some preparation to be able to recognize such things.

Vieten: Yes, I see you have read Dean's Wikipedia page.

Bengston: I have never read a Wikipedia page. I am sorry.

Vieten: Just kidding. Okay, we have enthusiasm from the audience. How about Jin Choon Kim?

Jin Choon Kim: I have two quick questions. One is, what is the origin of healing energy? I think it is different from gravitational energy and electromagnetic energy. Maybe it has some relationship with zero-point energy. In the whole universe, zero-point energy is different from gravitational energy and electromagnetic energy. So, I would like to hear Dr. Bengston's opinion about that. Second, is there any way how to explain the process of cognition or healing? In other words, maybe we could try to understand it through interactions between the consciousness field and the quantum information field of healers, in brain cells. So, I would like to hear your opinion about the quantum process of healing between healer and healee, especially between the consciousness field and the information field in the brain.

Bengston: I would give almost anything for the answer to your first question. I can tell you, I have not been particularly successful. So, whether healing has something to do with zero-point fields, or this or that, I simply do not know. I can tell you, though, empirically, that healing is not a simple field effect. So, however you end up with that interpretation, I am not sure where it is going to go. Let us say I have a cage of mice that I am treating here. Also, I have other mice at a distance, that are part of my experiment. Between these cages and those cages, there are other cages of the same mice. We can fix the distant mice. You would expect that the ones in between should experience some sort of effect, but they do not. So, the only way that they are affected is if they have been in some way or other identified—by some consciousness someplace, maybe the consciousness of the mice, but they are identified at someplace as being within the experiment. If you are within the experiment, the distance does not matter. But going a distance does not create that much of a larger field. So, I cannot speak to whether zero-point energy is related. I can say it is not a field effect. There is something else going on. That is probably not very satisfying to you, but it is the best I have.

Vieten: Dean?

Radin: I would add, though, that Bill, you have used the idea of entanglement as a way of thinking about these things. There is a similarity there that if in entanglement, whether it is particles or even larger systems, there has to be a relationship created. Then it is a system, it is created as a full, holistic system. So, you could have a context where some mice are simply not part of the definition of the whole, and then they are not entangled—so they do not show an effect.

Bengston: They do not become participants in the bonded group.

Radin: Yes. We do not know whether we are talking about it as a metaphor at this point, or whether there is some super-entanglement phenomenon that we do not know yet, or something like that.

Bengston: Yes, a good number of these wonderful questions are past my paygrade.

Vieten: How about Andrew Wilson?

Andrew Wilson: My question is in a similar vein of looking for some kind of theoretical basis for talking about this phenomenon. I absolutely agree with Dr. Bengston that consciousness or awareness is not a good descriptor. But I am also aware that looking for the ultimate cause in some kind of physics that does not yet exist, like zero-point energy, may be going a bit too far. But my preference is just to say that there should be some kind of a psychic analog to the physical world. Just as there is a world of humans, there is a world of spirits, an earthly world and a spirit world—as above, so below. So, why not consider that healing energy or the medium for ESP (or whatever) is some kind of energy that operates by some of the same principles that physical energy does. For instance, it has a frequency, there might be resonance, there might be some kind of giving and receiving interaction that could be strong or weak, depending on the degree of resonance between A and B. Are there people who are trying to work on that on a bit more modest level, to develop some kind of theory about healing or psychic energy?

Bengston: I think there is a whole lot of people with all sorts of speculative ideas about what this thing might be. One of my problems is that I do not know of any that you could not generate a whole lot of counter hypotheses for. I do not have the conceptual tools to wrap myself around this in terms of an explanatory and predictive model. I just do not. It could be staring me in the face, but I cannot unravel it. Nothing has clicked. Frankly, it is in that vein that I have started to move toward very practical applications—can we store it, bottle it, and scale it. Healing is interesting as a phenomenon in and of itself. But let us say you can heal, and someone else can heal, and let us even say it can be taught. Then you have a progression of people who know how to heal. There is a virtually inexhaustible supply of people in pain out there, and unless you can scale it, it is never going to be made conventional. I think it needs to be disentangled from a lot of the speculation.

Vieten: Okay, let us go to Thomas McDevitt.

Thomas McDevitt: Thank you for these fascinating presentations, and congratulations for launching this conference. My question concerns the socio-political implications of what you are discussing here. In your view, does the postmaterialist view of reality, as generally expressed in your findings, have implications that would challenge the materialistic view of human life that is so central to authoritarian regimes? Could "follow the science"—a popular cliche today because of COVID-19—have an impact sometime in the future, on the geopolitical conflict between East and West? Or, for example, between the communist view of life and the free world, to use those terms? As Chairman of *The Washington Times,* I am absolutely fascinated with this question that has real news value in terms of what you all are discovering. In that context, the primary issue today facing the entire world is the conflict between a communist view of life and a free-world view of life, generally speaking. I hope that makes sense. Thank you.

Vieten: Yes, Dean.

Radin: I think it does have very significant consequences for the way that civilization works. For example, from a materialistic perspective, there is the assumption of isolation, the assumption that the way all of science works, especially with the control design, you assume that a system can be isolated and is separate. And we often treat each other as isolated and separate. But among other things, the kinds of phenomena we are talking about, whether it is telepathy or some other phenomenon, suggest that there is no separateness. Second, at that level, there are no secrets. So, imagine a political system where there are no secrets—it would not look like the world we have today. In fact, politics could not work the way it does. Hardly anything would work the way it does. Government would be completely different. So, this is one of the consequences, and I think it may be one of the reasons why there is pressure to not exactly suppress, but to keep these kinds of ideas on the fringe—because the consequences of changing a worldview even a little bit toward this direction would have very significant consequences for the way the world works.

Vieten: Thank you. Do we have one more rapid, wrap-up comment from one of our speakers or panelists?

Bengston: I completely agree with Dean on that, you know. He said it better than I can say it. If you took a look at it from the perspective of healing, healing is connection, and, I daresay, healing is love.

Vieten: Thank you. Let us wrap up with Dr. Bahng.

Bahng: I would like to comment on healing energy, which may be helpful to understand it. I had a discussion about spoon bending—I think everybody knows about spoon bending. I asked the person, "How did you bend the spoon with your mind?" The reply was, "I just recall the memory of the spoon when it was at a high temperature; just by recalling the memory, the spoon would bend." So, it may be understood that it happened based on the information. In terms of healing energy, although we try a physical approach to understand it, I think we also have to consider the information itself.

Vieten: Thank you very much. Thank you to the speakers and commentators—we really appreciate your presentations. And thank you to the conference organizers. ◇

ICSG II

SESSION 2

The Promise of New Paradigm Science in Solving Environmental Problems

About Session 2

The Promise of New Paradigm Science in Solving Environmental Problems

Scientists and inventors using new paradigm science approaches have devised groundbreaking methods and technologies that offer a range of potential applications, such as for the diagnosis and treatment of diseases and for the protection and restoration of the natural environment. The presentations in this session cover two different but significant areas. The first presentation provides scientific evidence that we are not individual entities separate from one another and from our surroundings. Rather, we are deeply interconnected with other humans and with the environment. This realization should motivate us to protect nature and our fellow human beings. The second talk is on biohybridization, which combines conventional and unconventional areas of science. This talk describes a newly developing approach that integrates innovative technologies, such as artificial intelligence, with living systems. This approach has a wide range of potential applications, including prosthetic devices and implants for medical purposes and biosensors to monitor the conditions in ecosystems. ◇

About the Chair, Professor Harald Walach

- Founding Director, Change Health Science Institute, Berlin, Germany
- Visiting Professor, Department Psychologie, Universität Witten-Herdecke in Whitten, Germany
- Until July 2021, Professor, Poznan Medical University, Poland
- Until 2016, Professor, Viadrina European University, Frankfurt, Germany
- Author of more than 200 peer-reviewed papers, more than 100 book chapters and 16 books. Editor of the book series *Neuroscience, Consciousness and Spirituality*. His most recent book is *Bridges between Psychotherapy and Spirituality* (in German)
- His interest is in the interface of consciousness, health and spirituality within a wider scientific framework
- PhD in History and Theory of Science (2005, University of Vienna), PhD in Clinical Psychology (2001, University Bale)

Session 2 | Presentation One

Momentous Applications from the Frontiers of Science

Professor Arnaud Delorme
CerCo Laboratory, CNRS,
Paul Sabatier University, Toulouse, France

The topic of my presentation is, "Momentous Applications from the Frontiers of Science." I will begin by proposing the scientific hypothesis that consciousness is a fundamental property of nature (**Figure 1**). The idea is to use the tools of science to test this hypothesis.

Science is the "religion" of the twenty-first century. If science says something is true, then it becomes true for many people, and they start acting accordingly. What are the consequences and applications of frontiers-of-science research? Under this model, we are not separate from each other and our environment. Only when we realize this, at a deep organic level, will we come to respect each other and our environment genuinely.

In this presentation, I will first review some experiments showing our

About the Speaker

- Professor, CerCo Laboratory, CNRS, Paul Sabatier University, Toulouse, France
- Faculty, Swartz Center for Computational Neuroscience, University of California, San Diego, USA
- Scientist, Institute of Noetic Sciences, Petaluma, California
- Co-Principal Investigator, EEGLAB: Software for Analysis of Human Brain Dynamics
- Primary research interest is in analysis and modeling of human consciousness as captured by high-dimensional electroencephalography (EEG) and other imaging modalities
- PhD in Neuroscience; Investigating how technology could be used to assist meditation practice

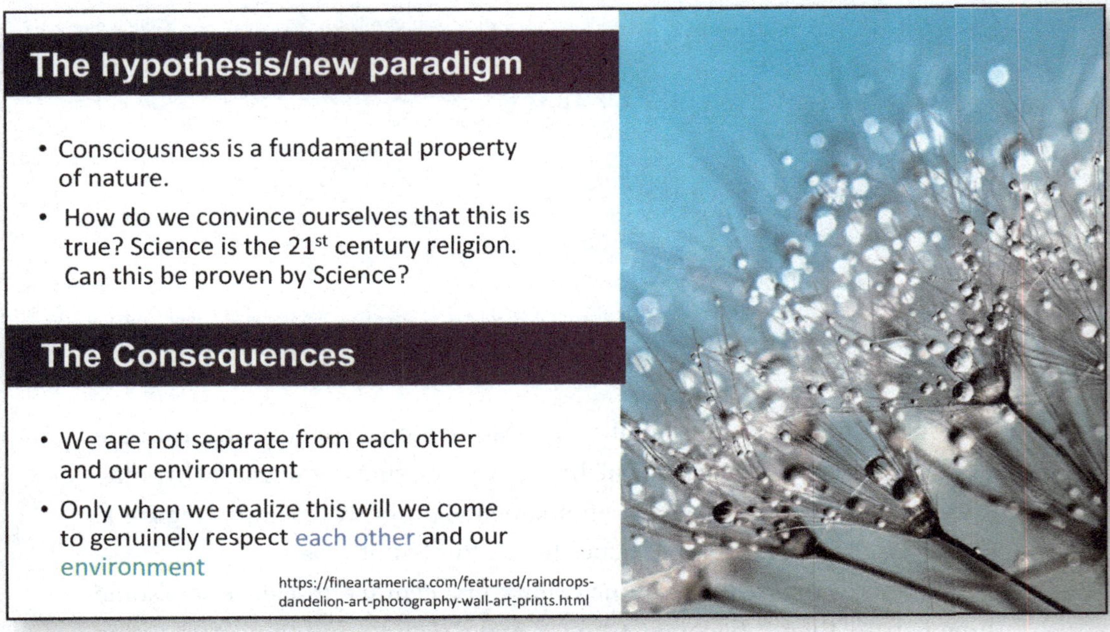

Figure 1

deep connection with our environment, then proceed with my own research showing our deep connection with each other.

First, our brain and body are connected to Earth's magnetic field. The perception of the geomagnetic field is well-established in various animal species. Still, its presence in humans has rarely been tested, yielding inconclusive results.

Recently, a scientific report showed a strong, specific human brain response to Earth-strength magnetic fields (**Figure 2**). The bar plot shows alpha brain activity in response to a real change in magnetic field as opposed to sham in blinded conditions. This research shows that we are connected with Earth's magnetic field, and I believe it will lead us to feel less separated from our environment and come to respect it more.

Figure 3 shows how our body is connected to cycles of the Moon. Many species synchronize reproductive behavior with a particular phase of the lunar cycle to increase reproductive success. The human menstrual cycle has a period close to that of the lunar cycle, and a recent paper showed the

Our body and brain feel the earth (magnetic field)

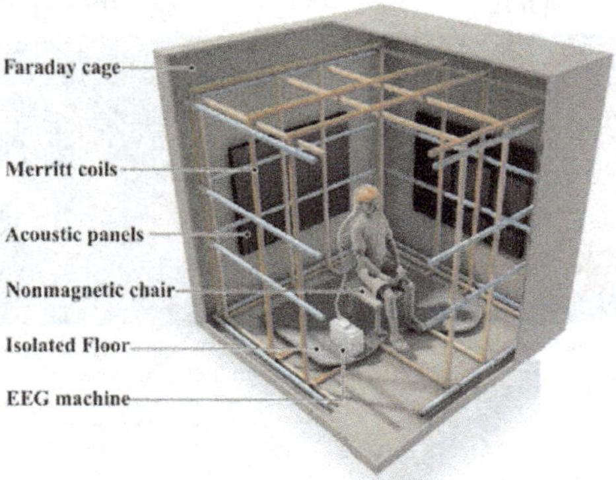

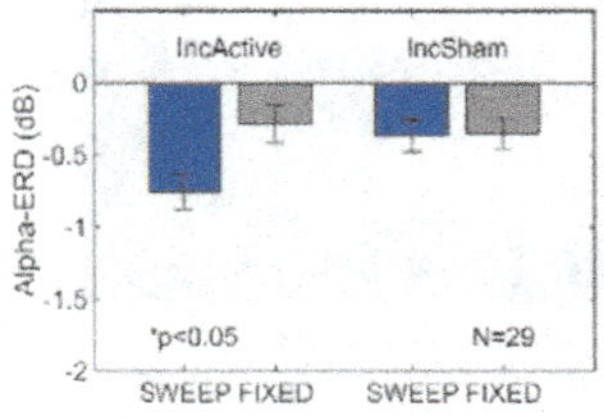

Consequence for solving environment problems: showing people that we are entangled with the earth "breath" and activity will lead to more respect of the environment.

Wang CX, Hilburn IA, Wu DA, et al. Transduction of the Geomagnetic Field as Evidenced from alpha-Band Activity in the Human Brain. *eNeuro*. 2019;6(2):ENEURO.0483-18.2019. Published 2019 Apr 26. doi:10.1523/ENEURO.0483-18.2019

Figure 2

Our body responds to planetary/solar changes

Women's menstrual cycle and moon phases

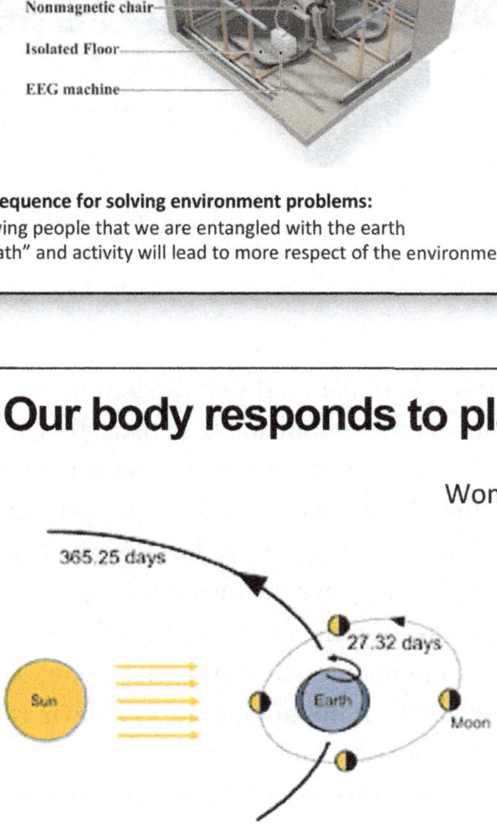

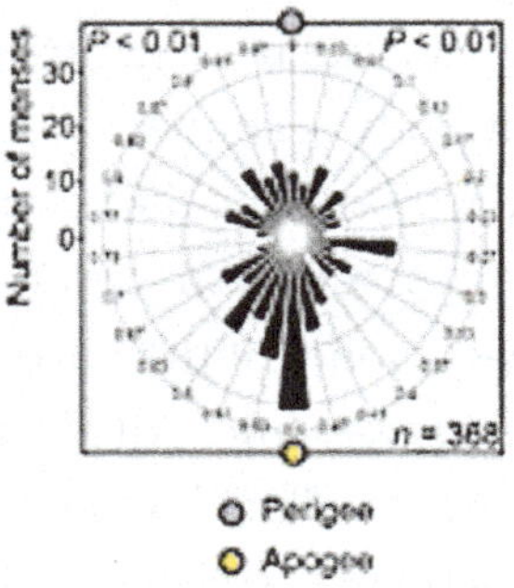

Demonstrating we are dependent on planetary activity will lead to the realization we are not independent of our environment. It will lead to less selfish behavior, more respect for the environment.

Helfrich-Förster C, Monecke S, Spiousas I, Hovestadt T, Mitesser O, Wehr TA. Women temporarily synchronize their menstrual cycles with the luminance and gravimetric cycles of the Moon. Sci Adv. 2021 Jan 27;7(5):eabe1358. doi: 10.1126/sciadv.abe1358. PMID: 33571128; PMCID: PMC7840133.

Figure 3

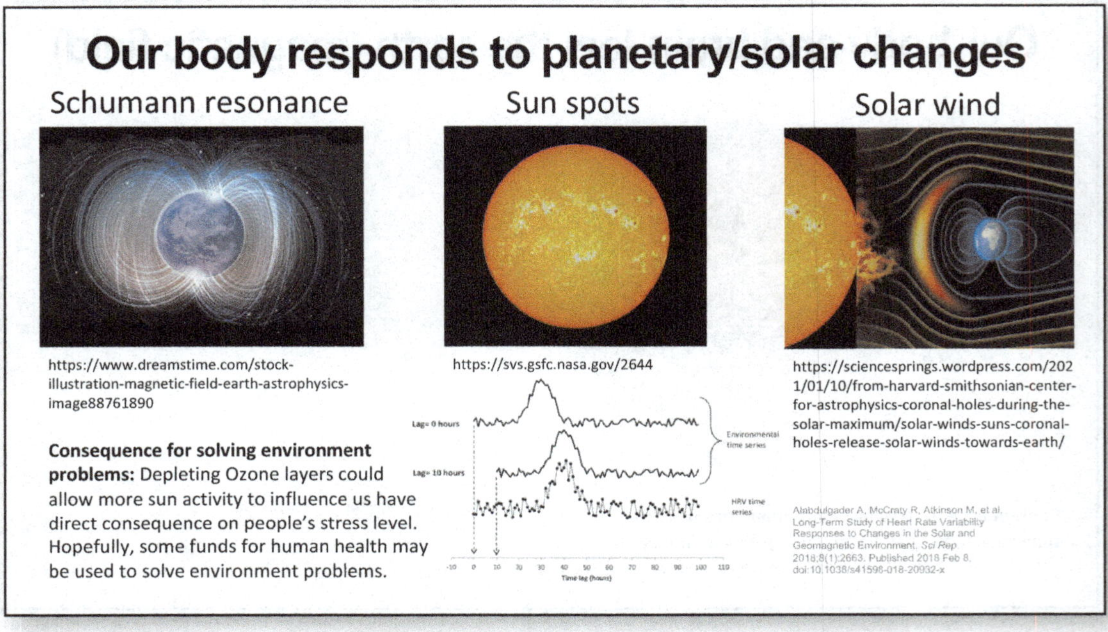

Figure 4

lunar influence on reproductive behavior in humans. The study showed that the period of menstruation in humans is not random and is statistically synchronized with the Moon cycle. As noted for the previous experiments, showing that we are connected with the Moon cycle will reduce our feelings of being separate from nature and enhance our respect for the environment.

Our body is connected to other planetary events, including activities on the Earth and the Sun. A recent study examined relationships between solar and magnetic factors and autonomic nervous system responses (**Figure 4**). Some of the Earth's magnetic field fluctuations, also known as *Schumann resonance,* the number of sunspots, and the amplitude of the solar wind can affect human psychophysical processes. This shows that our stress levels are correlated with our planetary and solar events. Depletion of the ozone layer may not only increase the risk of skin cancer but also increase the influence of the Sun's activity on people's stress levels. Because of this study, one may hope that some public funds currently allocated to

Credit: Courtesy of HeartMath Institute

Figure 5

improving human health may be reallocated to solve environmental problems and so indirectly help protect human health.

Trees and plants are also dependent on the Earth's activity. This is an emerging area of research. For several years, the HeartMath Institute (HMI, heartmath.org) has studied the electrical activity of trees in relation to our interconnectivity with nature. They measured this activity from individual trees as shown in **Figure 5**. They found that trees have a surprisingly complex range of electrical activity and rhythms, including electrical responses to earthquakes. These are preliminary results that have not yet been peer-reviewed, but they look promising.

Now HeartMath has a new project to study groups of trees. By looking at the activity of trees in groves at different locations in a forest or in urban environments, they hope to see how trees may be communicating with each other and how they may respond as a group to human emotions. Since both the EEGs (electroencephalograms) of people and the electromagnetic activity of trees seem correlated with Earth's magnetic field, this might be one

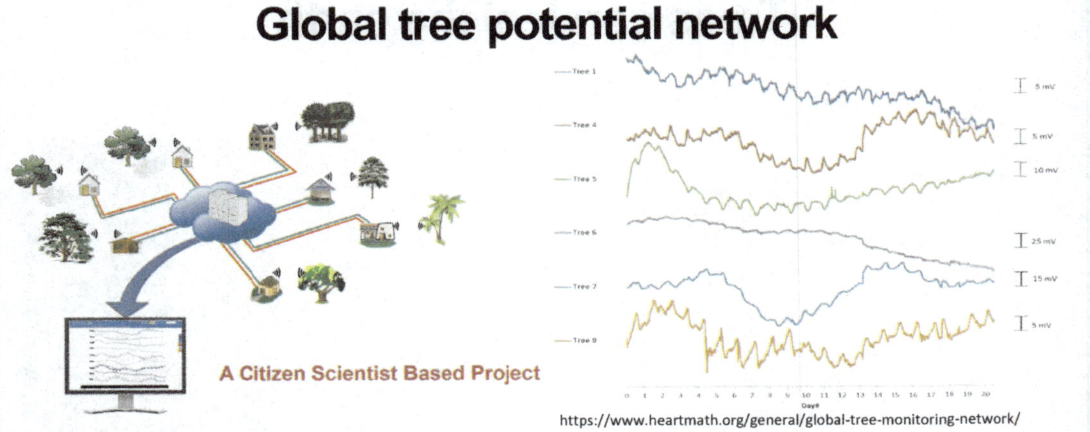

Consequence for solving environment problems: both people's EEG and tree's electro-magnetic activity seem correlated with earth magnetic field. This might be one of the scientific basis as to why we feel at peace in nature. As more tangible proof of our direct connection with nature are shown, we will come to respect nature.

Figure 6

scientific basis for why we feel connected with and at peace in nature (**Figure 6**). As additional tangible proofs of our direct connection with nature are shown, we will come to respect it increasingly.

I will now review experiments I have done showing that human consciousness might not be in the body, again demonstrating that we are not separate entities independent of our environment. Instead, we are deeply connected with all things and with one another.

Figure 7 shows the results of an online survey that some of my colleagues and I conducted to investigate the prevalence of a wide range of experiences in more than 1,000 meditators. Most meditators feel a connection to others and their environment that transcends the materialist paradigm, including the experience of telepathy, for example. Our results show that most respondents report they had many anomalous and extraordinary experiences. These are subjective reports, so they do not demonstrate that these experiences are real, but at least they warrant further investigation of this topic.

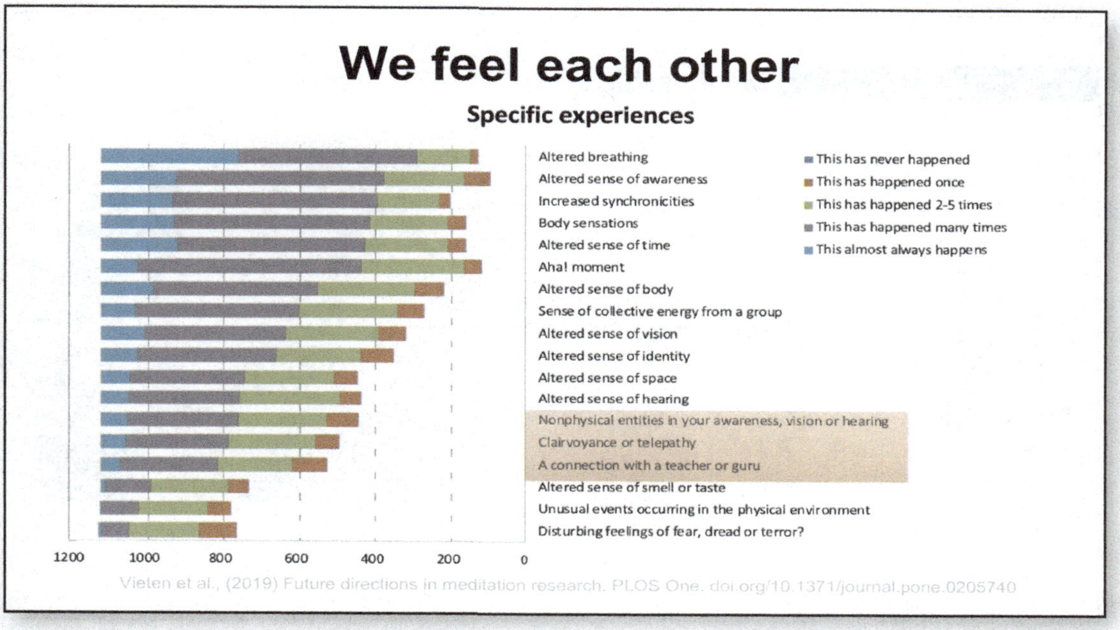

Figure 7

Demonstrating our interconnectedness Mediumship/Channeling, why?

- Psychics and Mediumship readings are the only recognized "paranormal" job!
- In comparison to Psychics, Mediums are (relatively) easy to test
- Extreme emotions – such as the ones associated with death – usually trigger the most spontaneous paranormal reports
- There are prior reputable studies:

  Beischel, J., & Schwartz, G. E. (2007). Anomalous information reception by research mediums demonstrated using a novel triple-blind protocol. *EXPLORE: The Journal Science & Healing*, 3, 23–27.

  Kelly and Arcangel (2011) An Investigation of Mediums Who Claim to Give Information About Deceased Persons. *The Journal of Nervous and Mental Disease*. Volume 199, Number 1.
- Simple hypothesis: there is something going on in Mediumship the current scientific model fails to explain — could generalize to other psi effects

Figure 8

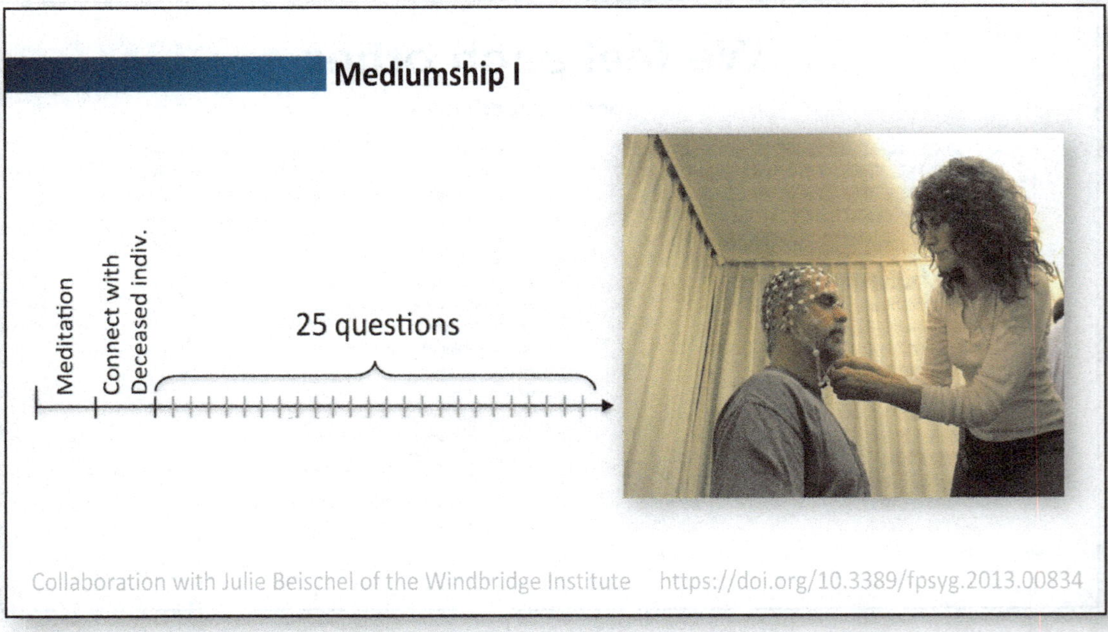

Figure 9

Thus, I decided to study mediumship, which is an ancestral practice in which people with this gift access information they are otherwise not privy to. It is an ideal candidate for testing the hypothesis that consciousness is nonlocal to the brain. First, it is widely practiced. Second, it involves extreme emotions—such as the ones associated with death—which usually trigger the most spontaneous paranormal reports. There are also prior reputable studies on the accuracy of mediums in blinded conditions, and two such investigations are noteworthy (**Figure 8**).

My hypothesis is simple here. The goal is not to show that we survive after death. It could just be telepathy between medium and client. My hypothesis is simply to test if there is something that the current materialistic scientific model fails to explain.

Therefore, I am going to briefly detail three experiments we did on this topic. In the first, we collected behavioral and electro-encephalography data from several mediums under double-blind conditions. Mediums

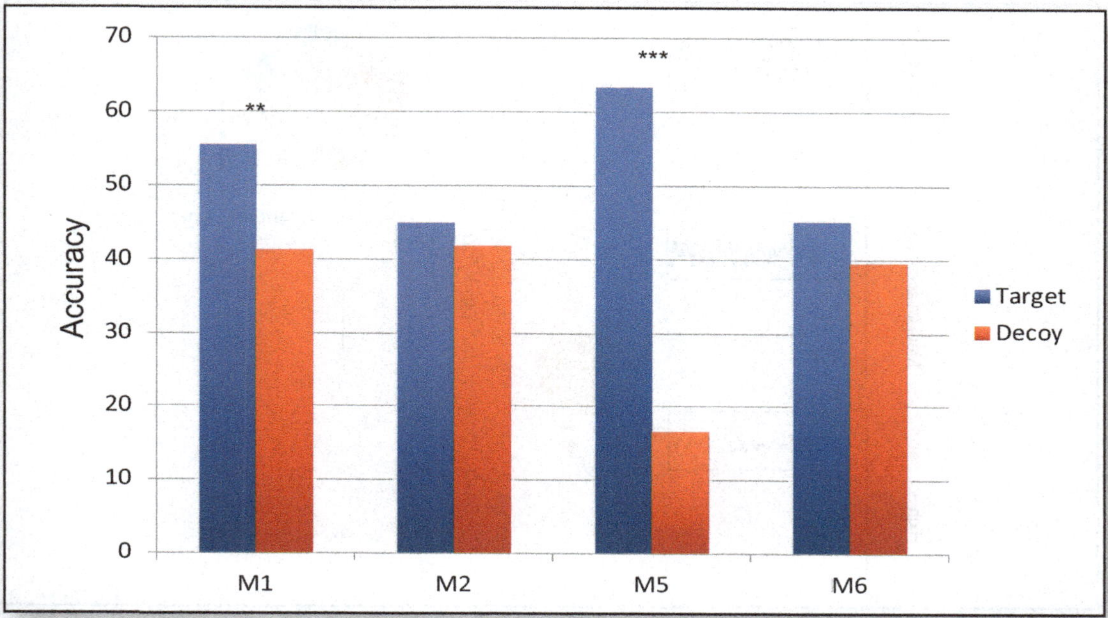

Source: Delorme et al. 2013.

Figure 10

were given only the first name of a deceased person and asked 25 questions about that person (**Figure 9**).

The responses to these questions were then rated for accuracy (**Figure 10**). Mediums' accuracy was rated above chance expectation (blue bars) when compared to control data for each medium (red bars).

We then looked at the mediums' brainwaves. **Figure 11** shows the results for one of the mediums. We observed a correlation between accuracy and theta (so 5 Hz) brainwaves—the smaller the theta activity, the more accurate this medium was. Theta activity, especially in this brain region, is linked to working memory. One interpretation was that the medium was emptying her working memory and becoming more receptive to receiving accurate information.

These results were published in a mainstream journal (Delorme et al. 2013) and gathered attention from the scientific community. These differences suggest that mediums had access to nonlocal information, which was

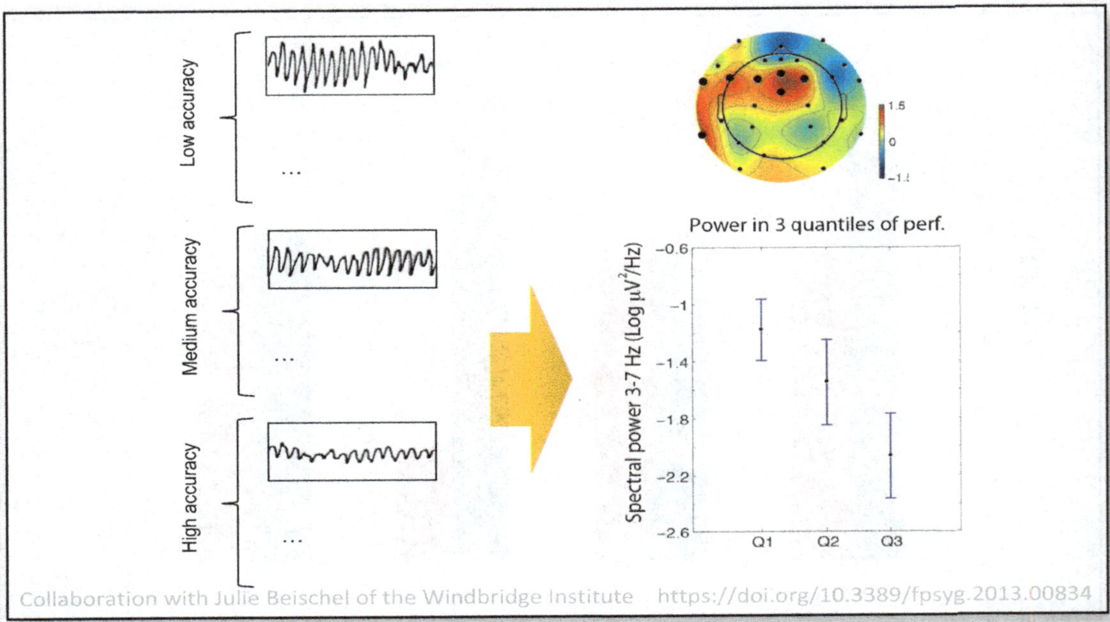

Figure 11

Source: Delorme et al. 2013.

also reflected in the brain activity of some of them. This warranted further studies on the topic.

Our second experiment was structured with an improved experimental design to increase its statistical power (**Figure 12**). We invited 12 mediums to see if they could tell if a person was alive or dead based solely on briefly seeing a photograph of that person. Half of the individuals in the photos were deceased, and half were alive at the time the experiment was conducted. There were 404 photographs, shown one at a time, and the mediums had 8 seconds for each to indicate if they felt the person was alive or deceased.

The images of the living and deceased individuals were balanced using 11 image characteristics so that no single characteristic could help with classifying the images. For example, we would not want to have all the photographs of the living people smiling and the photographs of the deceased not smiling. We wrote software to balance, as well as possible, the image characteristics. This involved, as shown in **Figure 13**, reframing the

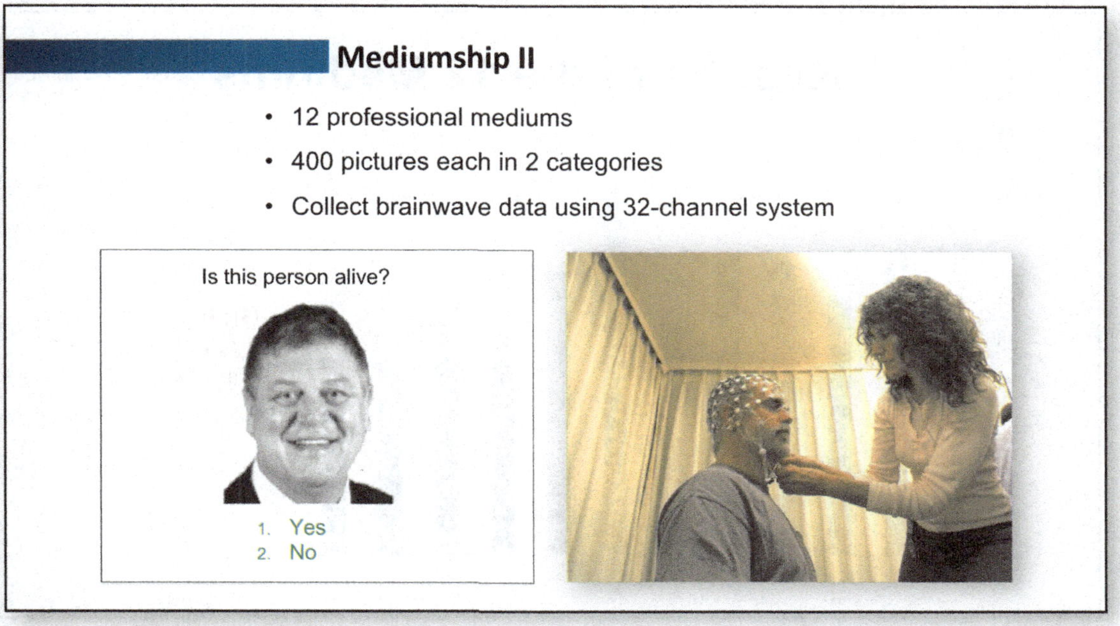

Source: Delorme et al. 2018.

Figure 12

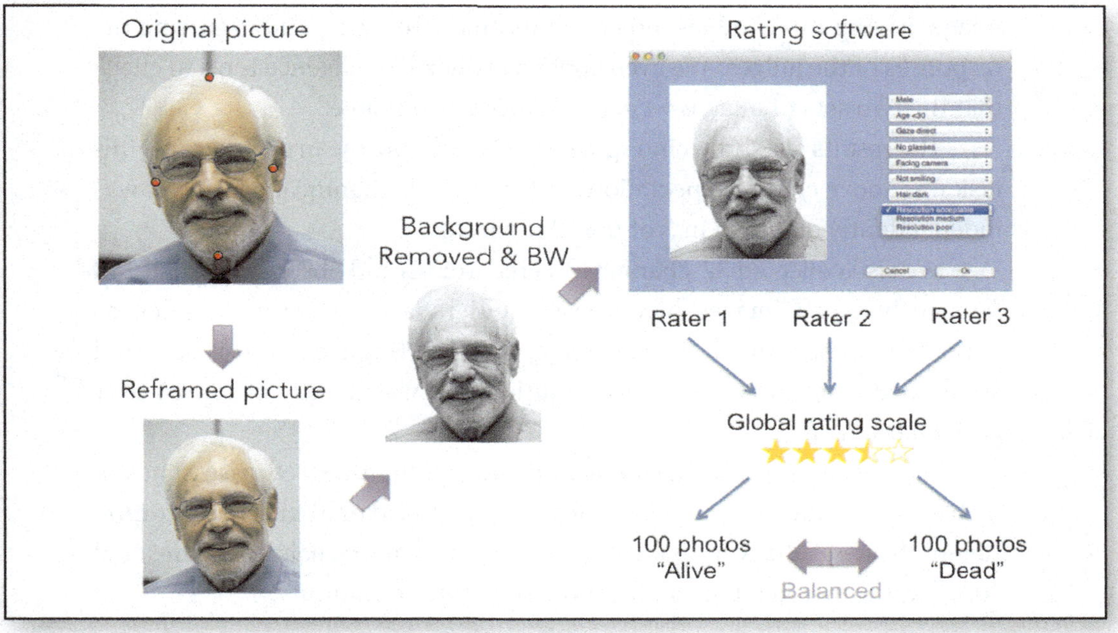

Source: Delorme et al. 2018.

Figure 13

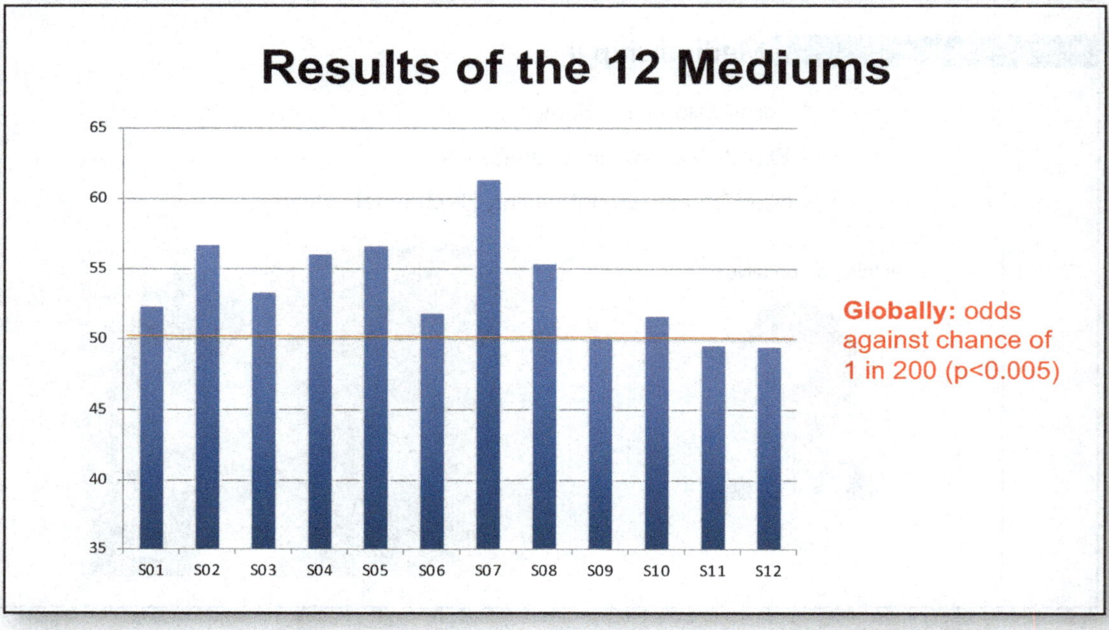

Figure 14 Source: Delorme et al. 2018.

image, having three judges indicating its characteristics, and balancing the responses of the judges. We even applied artificial intelligence tools to check that the two sets of images were as balanced as possible.

The results are shown in **Figure 14**. Overall, the mean accuracy for this task was above chance expectation, and statistically significant accuracy was independently obtained in 5 of the 12 mediums.

We also collected 32-channel EEG recordings and observed a robust difference between images of deceased individuals correctly classified as opposed to images that were incorrectly classified (**Figure 15**). This occurred at about 100 ms after image presentation and was mainly observed over parietal brain areas.

Our results suggest that some individuals can intuitively assess mortality based on some unknown facial features, or based on their pure intuition. Again, these results were published in mainstream journals (Delorme et al. 2018) and gathered attention from the scientific community.

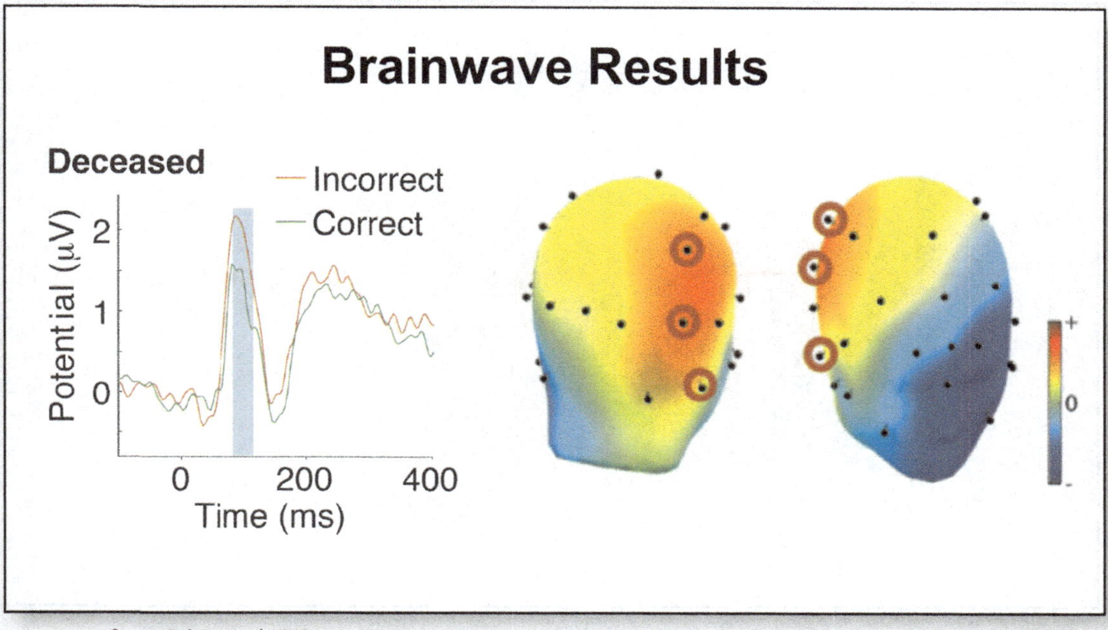

Source: Delorme et al. 2018.

Figure 15

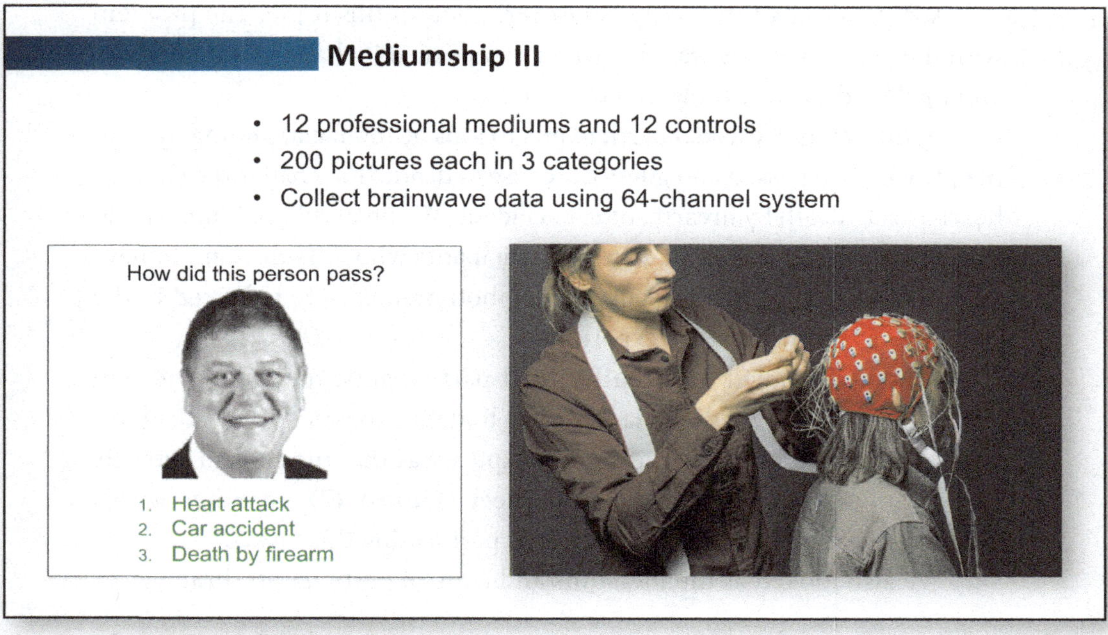

Source: Delorme et al. 2020.

Figure 16

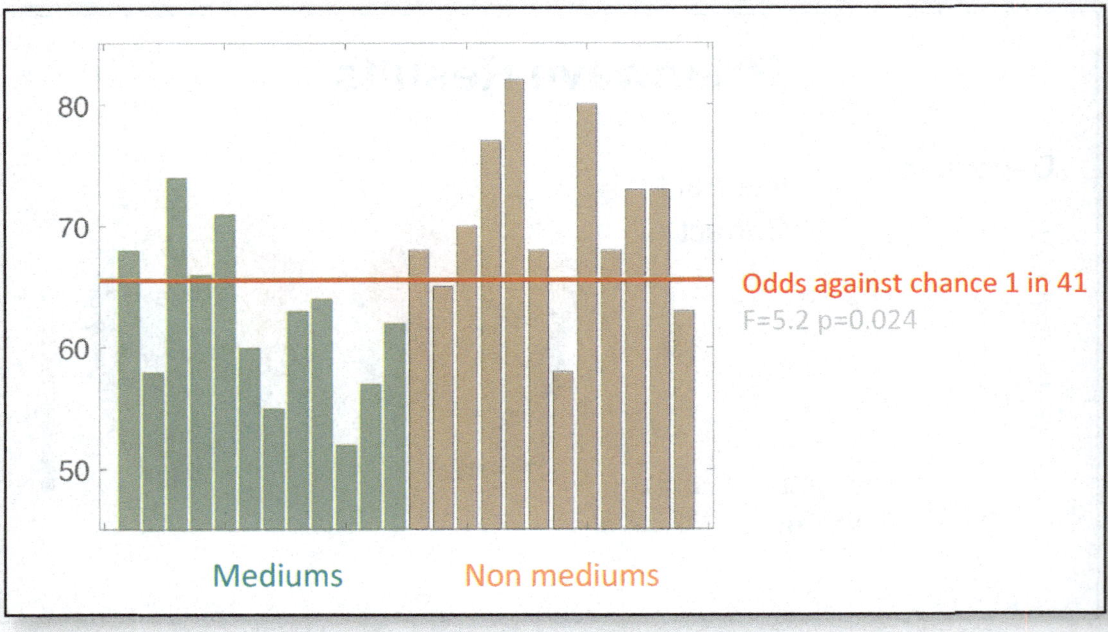

Figure 17

We then did a third and final experiment on this topic. The problem with the previous study was that when people died it changed our database and made it difficult to collect more data.

In this study, we asked participants to look at 180 facial photographs of people who had passed and guess the cause of death. They had three choices: heart attack, death by firearm, or car accident. We invited 12 mediums to do the experiment as well as 12 control participants who did not claim to have any mediumship abilities. Again, facial photographs were balanced in the same way as in the previous experiment.

When we pooled the data from all participants, mediums and non-mediums, we found that people were able to detect the cause of death above chance expectations. What we did not expect was that the control subjects were primarily responsible for this effect (**Figure 17**). In other words, mediums were no better than controls at performing this task.

We also looked at the mediums and control participants' brainwaves.

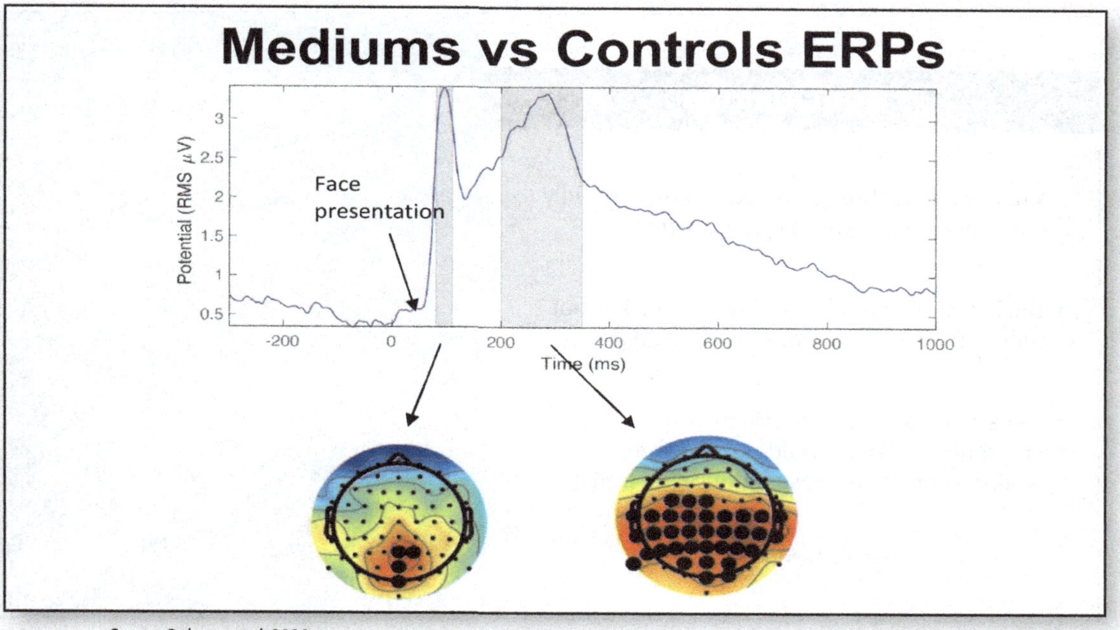

Source: Delorme et al. 2020.

Figure 18

We did not find the early difference we observed in the previous experiment, but it could be because the question we asked participants was different. We did find EEG differences between the mediums and controls. **Figure 18** shows these differences in neural activity, expressed in terms of event-related potentials (ERPs), which are time-locked EEG changes. The presentation of the facial photograph is at time 0, and it can be seen that the brains of mediums and controls react differently, with a significant difference at 100 ms after the image presentation and at about 300 ms after the image presentation. When looking at the literature as to what this difference could mean, it could be interpreted as reflecting greater attention and less inhibition by controls as compared to the mediums. It is possible that the mediums were stressed and had performance anxiety, while control participants did not.

This research was published in the mainstream journal *Brain and Cognition* and is one of our most popular papers (Delorme et al. 2020).

These results suggest that consciousness is not limited to the brain, and

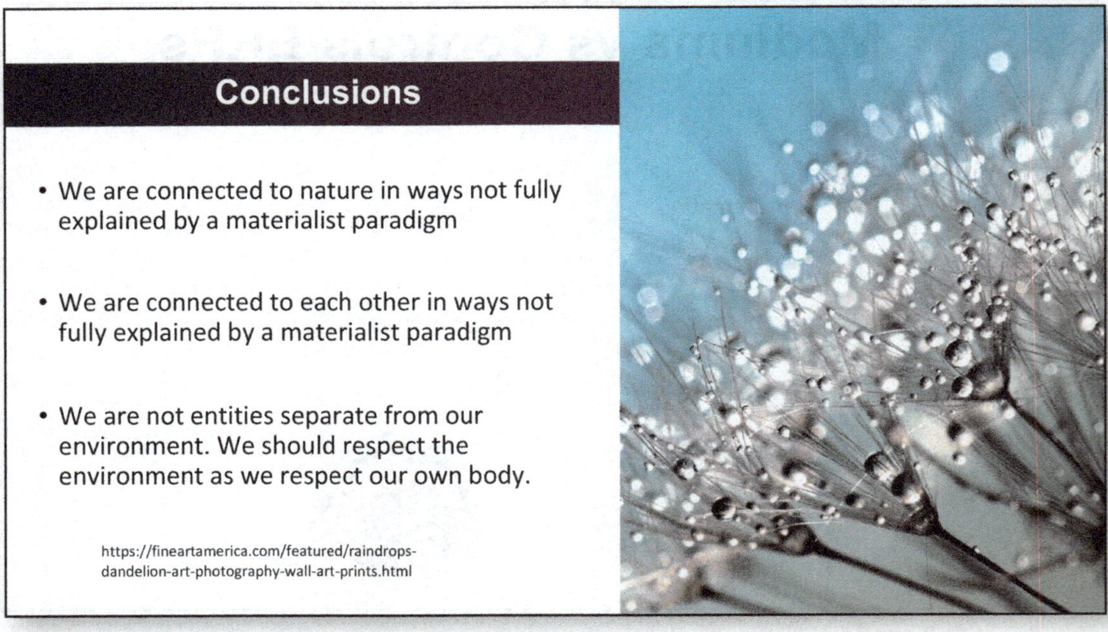

Figure 19

this could have important consequences for our relationship with the environment and how it may help solve environmental problems. First, there is a direct consequence of this research. If we can communicate with others, including plants and animals, directly using consciousness, we will come to respect them as part of ourselves. Second, there is an indirect consequence. Since mediums have been shown to help with bereaving, based on the work of Julie Beishel, for example, improved mental health for people receiving mediumship sessions may indirectly benefit the environment. As people experience less stress, they tend to resort less to coping mechanisms such as compulsive buying disorder, which contributes to a large amount of waste and harms the environment.

To conclude, how may we interpret the results of the studies presented here? Although these results do not yet demonstrate that consciousness is a fundamental property of nature, they indicate that this is a promising line of research. These results show, first, that we are connected to nature, including to the Moon, the Earth, the Sun, and maybe even trees, in ways we did not expect. Second, mediumship studies show we might be connected to each other in ways not fully explained by the materialist paradigm. We are not entities separate from our environment. We should therefore respect the environment as we respect our own body (**Figure 19**).

Thank you for your attention. I also want to thank everyone who helped make this research possible, including the BIAL Foundation and my collaborators and students. ◊

References

Alabdulgader, A., R. McCraty, M. Atkinson, Y. Dobyns, A. Vainoras, M. Ragulskis, and V. Stolc. 2018. "Long-Term Study of Heart Rate Variability Responses to Changes in the Solar and Geomagnetic Environment." *Scientific Reports* 8 (1): 2663. https://doi.org/10.1038/s41598-018-20932-x. Accessed on May 13, 2022.

Beischel, J., and G.E. Schwartz. 2007. "Anomalous Information Reception by Research Mediums Demonstrated Using a Novel Triple-Blind Protocol." *EXPLORE* 3 (1): 23–27. https://doi.org/10.1016/j.explore.2006.10.004. Accessed on May 13, 2022.

Delorme, A., J. Beischel, L. Michel, M. Boccuzzi, D. Radin, and P.J. Mills. 2013. "Electrocortical Activity Associated with Subjective Communication with the Deceased." *Frontiers in Psychology* 4: 834. doi: 10.3389/fpsyg.2013.00834. Accessed on May 13, 2022.

———, A. Pierce, L. Michel, and D. Radin. 2018. "Intuitive Assessment of Mortality based on Facial Characteristics: Behavioral, Electrocortical, and Machine Learning Analyses." *EXPLORE: The Journal of Science and Healing* 14 (4): 262–267. doi: 10.1016/j.explore.2017.10.011. Accessed on May 13, 2022.

———, C. Cannard, D. Radin, and H. Wahbeh. 2020. "Accuracy and neural correlates of blinded mediumship compared to controls on an image classification task." *Brain and Cognition,* 146, Article 105638. https://doi.org/10.1016/j.bandc.2020.105638. Accessed on July 1, 2022.

Helfrich-Förster, C., S. Monecke, I. Spiousas, T. Hovestadt, O. Mitesser, and T.A. Wehr. 2021. "Women temporarily synchronize their menstrual cycles with the luminance and gravimetric cycles of the Moon." *Science Advances* 7 (5). https://doi.org/10.1126/sciadv.abe1358. Accessed on May 13, 2022.

Kelly, E.W., and D. Arcangel. 2011. "An Investigation of Mediums Who Claim to Give Information About Deceased Persons." *The Journal of Nervous and Mental Disease* 199 (1): 11–17. https://doi.org/10.1097/nmd.0b013e31820439da. Accessed on May 13, 2022.

Vieten, C., H. Wahbeh, B.R. Cahn, K. MacLean, M. Estrada, P. Mills, M. Murphy, S. Shapiro, D. Radin, Z. Josipovic, D.E. Presti, M. Sapiro, J.C. Bays, P. Russell, D. Vago, F. Travis, R. Walsh, and A. Delorme. 2018. "Future directions in meditation research: Recommendations for expanding the field of contemplative science." *PLOS One* 13 (11): e0205740. https://doi.org/10.1371/journal.pone.0205740. Accessed on May 13, 2022.

Wang, C.X., I.A. Hilburn, D. Wu, Y. Mizuhara, C.P. Cousté, J.N.H. Abrahams, S.E. Bernstein, A. Matani, S. Shimojo, and J.L. Kirschvink. 2019. "Transduction of the Geomagnetic Field as Evidenced from alpha-Band Activity in the Human Brain." *eNeuro* 6 (2). https://doi.org/10.1523/ENEURO.0483-18.2019. Accessed on May 13, 2022.

Session 2 | First Commentary
On the Presentation by Professor Arnaud Delorme

The Three Dimensions of Time

Professor Jonathan W. Schooler
Department of Psychological and Brain Sciences,
UC Santa Barbara, USA

Greetings! It is a real pleasure to be speaking with you today. I want to commend Dr. Delorme on his provocative and intriguing talk. In science, most of us do not have the courage to investigate topics as controversial as the possibility that psychics might be able to access knowledge about the deceased. We would need unconventional models. Yet, Dr. Delorme has carefully and rigorously investigated this evidence and combined it with a variety of other sources of evidence to suggest that consciousness may exist in the physical universe in ways not currently appreciated by mainstream science.

Although I cannot say with great confidence that I believe necessarily any of the individual findings—particularly the ones about mediumship or at least the nonconventional account of them—that Dr. Delorme is suggesting, the possibility that consciousness may exist in ways not currently

About the Speaker

- Distinguished Professor, Department of Psychological and Brain Sciences, University of California Santa Barbara, USA
- Director, Center for Mindfulness and Human Potential
- Research areas: consciousness, memory, the relationship between language and thought, creativity, problem-solving, and decision-making
- PhD in Psychology, University of Washington

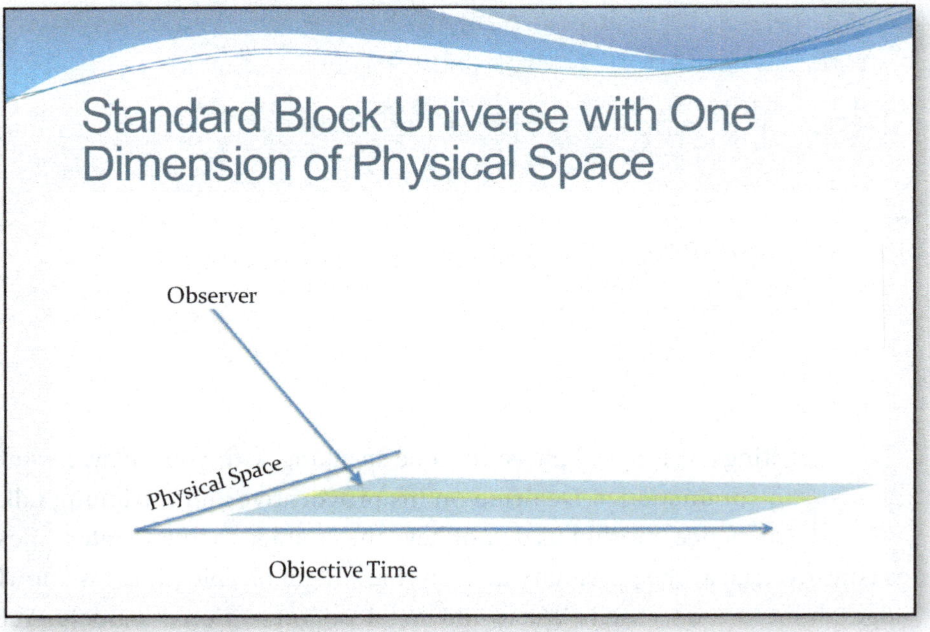

Figure 1

appreciated by mainstream science is very real to me, and I would like to flesh out that information with you right now.

I believe there are three self-evident truths that seem absolutely incontrovertible. First, experience exists. I am participating in an experiential state—even if that state is illusory, the experience itself exists. This is a notion pertaining to Descartes's view, "I think, therefore I am."

Second, experience is defined by change. Every moment I experience is not static; it is dynamic. The flow of time is an intrinsic aspect of my experience.

Third, experience is always in the now. I am never in the past and never in the future—I am always in the now.

The fascinating thing is that these three self-evident truths, the things that I know the best, are highly problematic for current conceptualizations of science. In fact, from the view of science, the flow of time is an illusion. Einstein famously said, "The past, present, and future are all illusions, even if stubborn ones." Physicist Paul Davies said, "Nothing in known physics

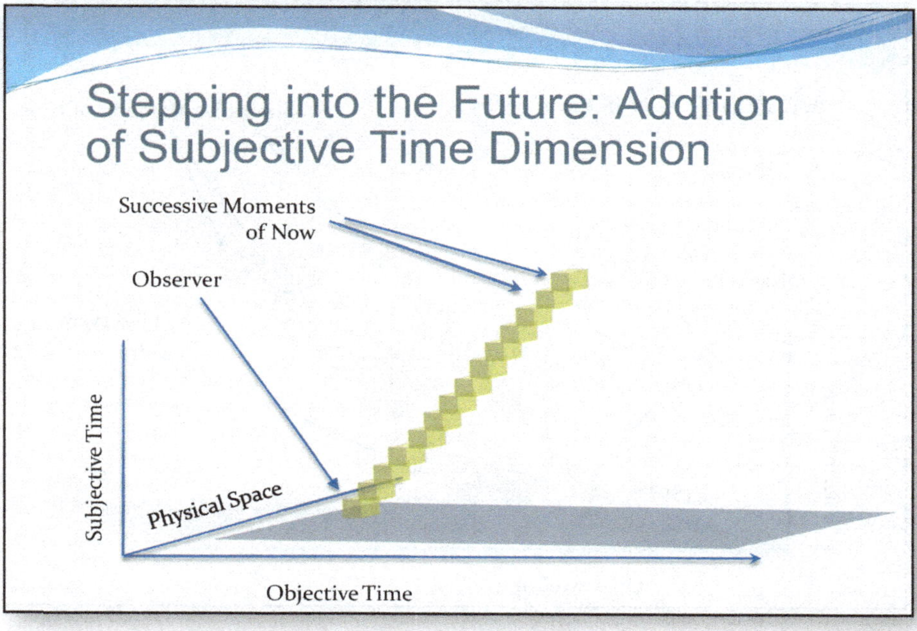

Figure 2

corresponds to the passage of time. Indeed, physicists insist that time doesn't flow at all; it merely is."

So here is the rub. The three things that I am most confident about, the existence of experience, the flow of time, and the privileged now all have no place in current views of physics, and, more generally, in science.

Why is that? I think the fundamental problem here has to do with something known as the block universe (**Figure 1**). Whatever else I may say, the key point is that we need to have humility to recognize that we do not understand things as they currently are, and we need to have a better understanding.

The problem is, we are stuck in a block universe where the past, the present, and the future all exist simultaneously. This leads us to the need for an extra degree of freedom, in order to be able to move through the block universe. So, I would like to posit the possibility of an additional dimension of reality: a subjective dimension, that is, a dimension of subjective time through moving in objective time relative to subjective time (**Figure 2**) (Schooler 2015).

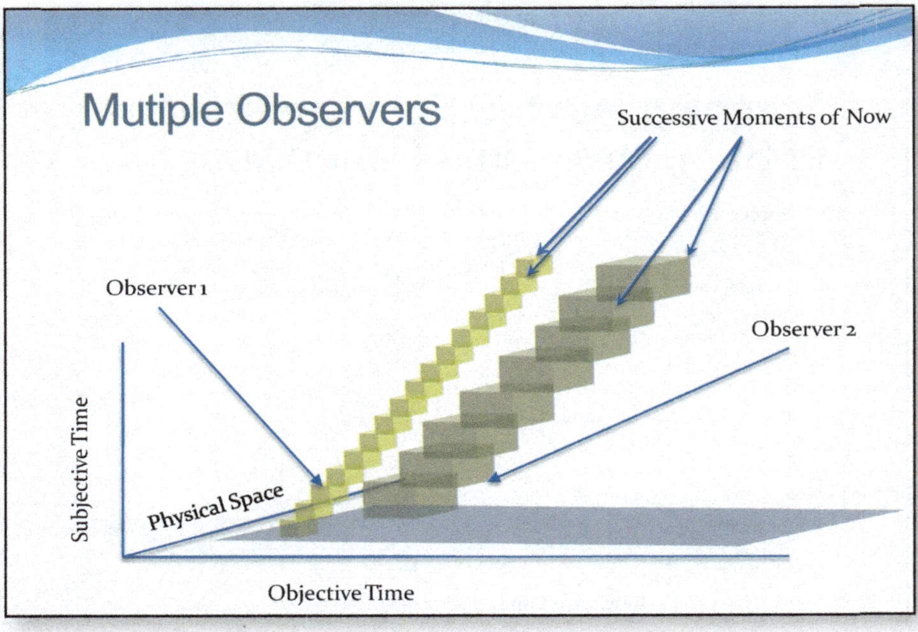

Figure 3

You can see this here. Here we have the block universe compressed. Where previously the block universe would compress the three dimensions to two, now I am compressing them into one. You can see that the observer is represented in all locations simultaneously—there is no movement. But when we introduce a subjective dimension, now the observer can move in subjective time relative to objective time.

It is possible that different observers are moving in objective time and subjective time at different rates (**Figure 3**). For example, a fly may be taking more steps per objective moment than we are. This may be why it is so hard to swat a fly—because, from the fly's perspective, in trying to swat it, you are going in very slow motion.

So, I suggest that we are moving in objective time relative to subjective time. Here we are, moving up; but in the same way, it could be said that we are in a static frame and time is coming at us. So, we are moving step by step by step, going up in subjective time and forward in objective time, as if we were going up the down escalator.

Now let me suggest an additional dimension of time: alternative time (**Figures 4, 5**). Physicists have proposed that we are constantly splitting off into additional, parallel universes, raising the possibility that when you are in one group, Schrödinger's cat is alive, and in another group, it is dead—we may be splitting off. This idea raises the possibility that there may not be a single future but multiple futures. In that case, we would be navigating between alternative possible futures. This raises the idea that we may be moving in objective time and subjective time and turning in alternative time.

Here is a vision of this model. We are going up in subjective time, objective time is coming at us, and we are turning left and right as we choose one possible alternative versus another. We may have minimal control over this, but perhaps even just our interpretations may drive that.

This further leads to the conjecture that there may be multiple levels of windows. So, we are at one window, but we may be looking at pixels upon pixels upon pixels, and there may be multiple windows moving through the dimensions of time, something I call *nested observer windows* (**Figures 6, 7**),

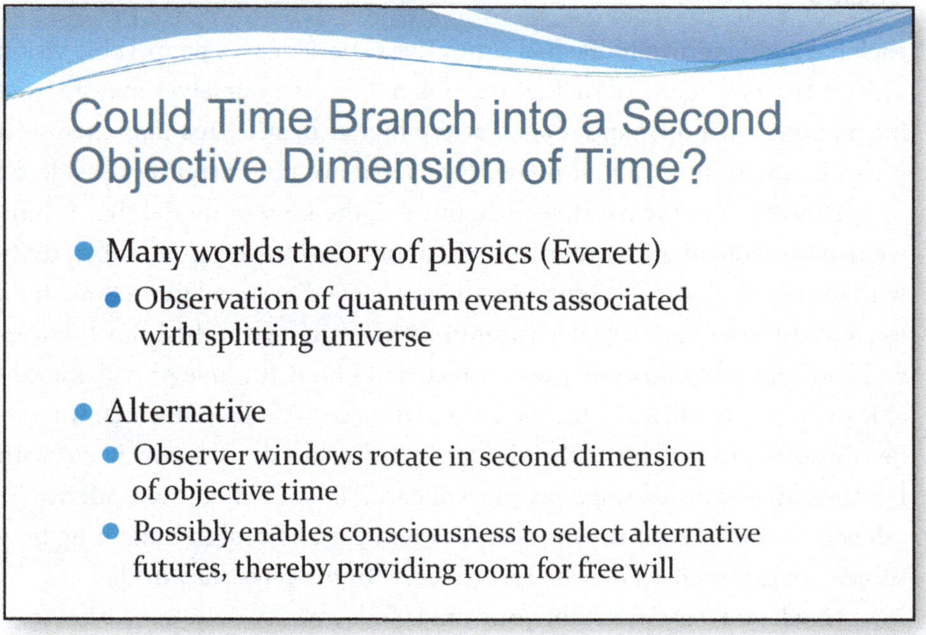

Figure 4

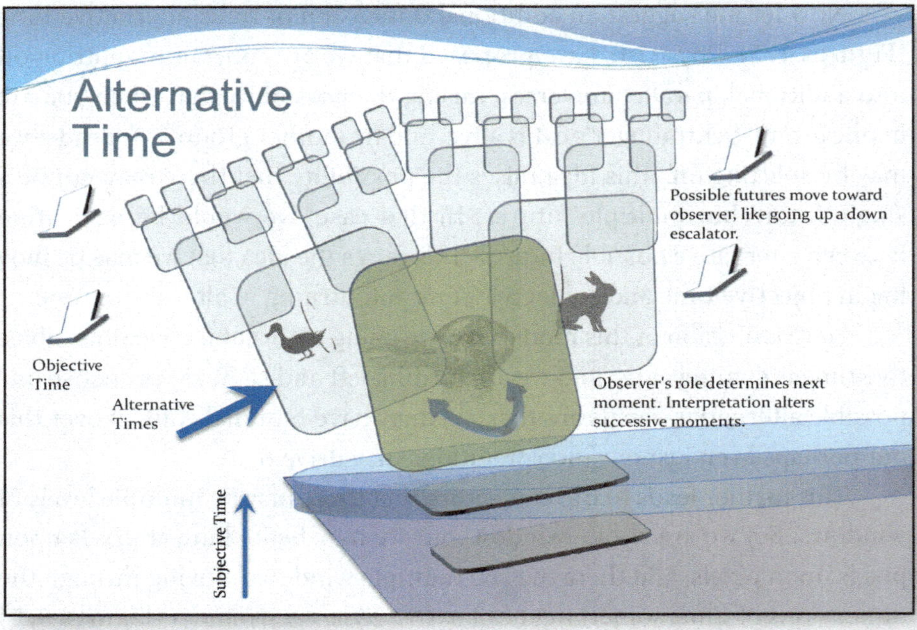

Figure 5

each having their own individual subjective experience—a panpsychic vision. This, then, may be the tiling that we look out on. We ourselves may be existing as both a window and a pixel at the higher levels. This may then be a possible way to think about higher levels that are not currently considered.

This is a very provocative idea, but it is the kind of model that I think we need to consider, in order to think about how to reconcile subjectivity with the physical world (**Figure 8**). I think we need to spend more time thinking about the metaphysical assumptions in which we embed our theories and consider other possible frameworks, and I hope this one provides a valuable one. In general, it invites the kind of thinking of a way in which we may be connected to the rest of the physical world that I think is consistent with Dr. Delorme's views. I hope people will carefully reflect on the evidence he offers and think about the models they may find helpful in imagining how subjective experiences may be integrated with the physical world.

Thank you very much for your time. ◊

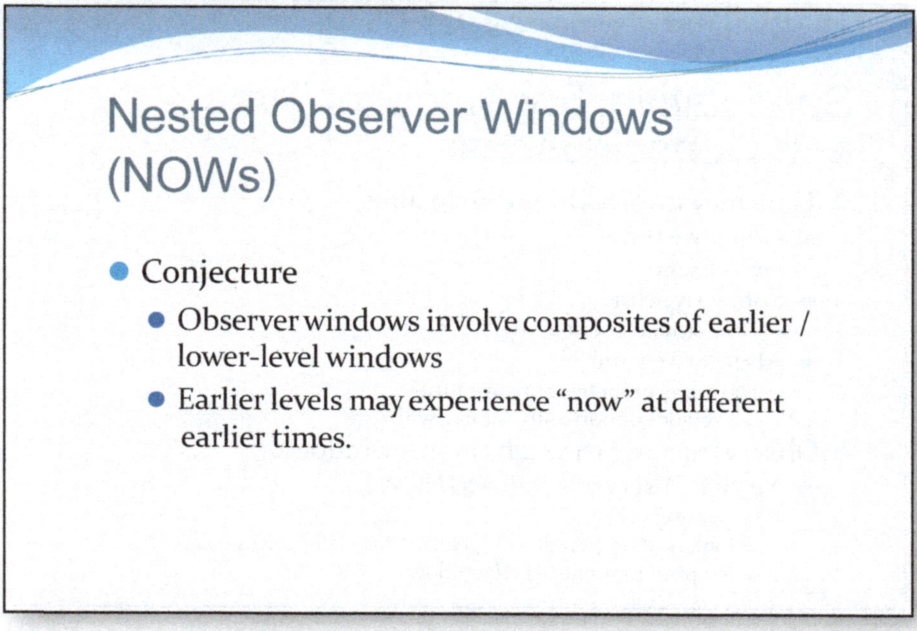

Figure 6

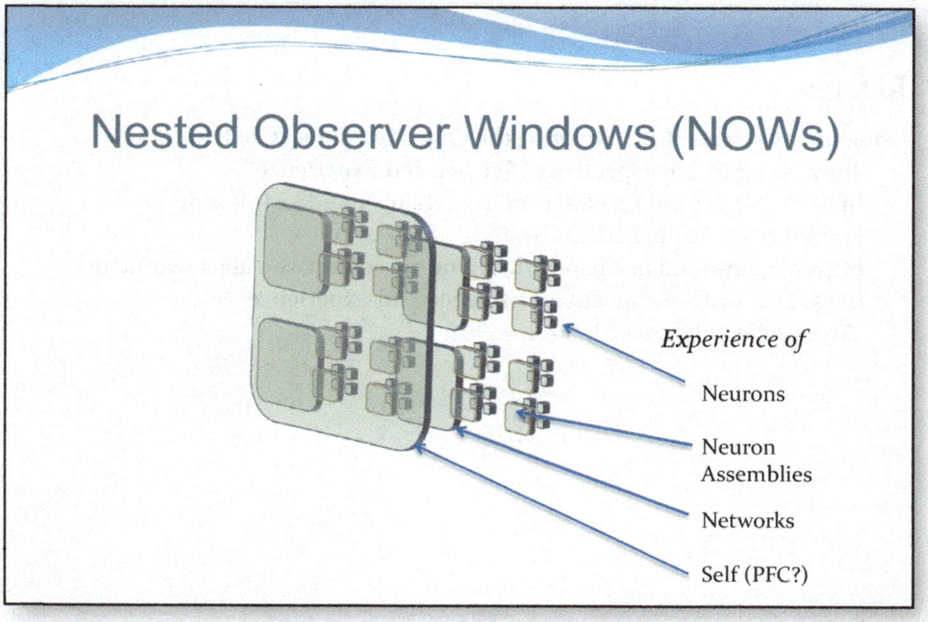

Figure 7

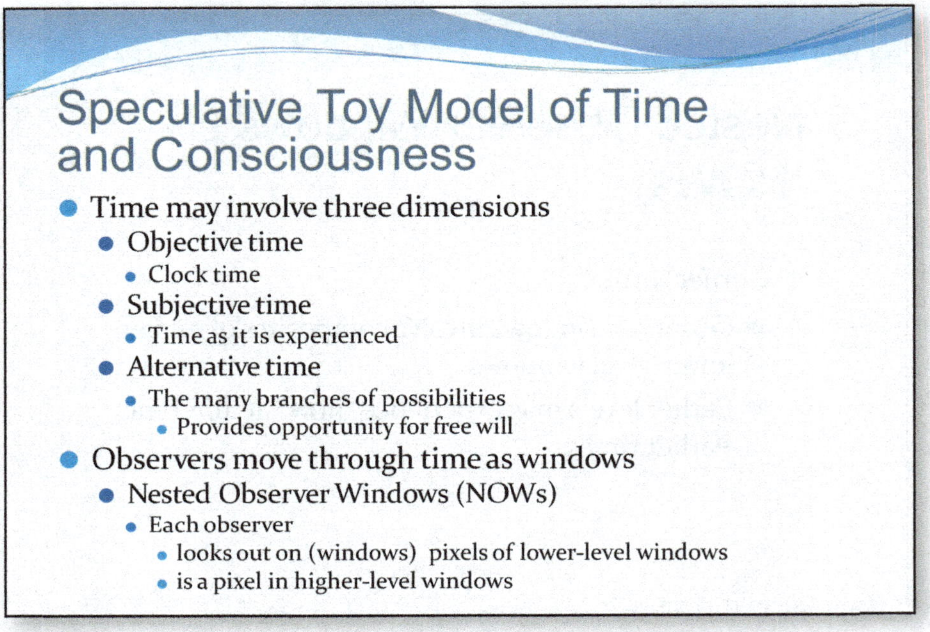

Figure 8

Reference

Schooler, Jonathan. 2015. "Bridging the Objective/Subjective Divide: Towards a Meta-Perspective of Science and Experience." In *Open MIND:* 34(T), edited by T. Metzinger and J.M. Windt. Frankfurt am Main: MIND Group. https://open-mind.net/papers/bridging-the-objective-subjective-divide-towards-a-meta-perspective-of-science-and-experience Accessed on October 31, 2021.

Session 2 | Second Commentary
On the Presentation by Professor Arnaud Delorme

Concerning Momentous Applications from the Frontiers of Science

Professor Won H. Kim
Department of Biochemistry, Yonsei University, Wonju, Republic of Korea

Dr. Arnaud Delorme presented various scientific experiments showing that we are not separate from one another and our environment, and our interconnectedness cannot be explained by the current materialist paradigm. Current science raises questions and answers only what can be explained based on materialism. If it cannot be explained by materialism, it is regarded as unscientific. Science that stays within the materialist paradigm can be compared to living in a well. Life inside a well

About the Speaker

- Professor, Department of Biochemistry, College of Medicine, Yonsei University, Republic of Korea
- Chairman, Society of Applied Subtle Energy (2014–2018)
- Editorial Board Member, *Journal of Vortex Science and Technology* (since 2014)
- Current research focuses on energy medicine and water science
- Inventor of revolutionary items based on new paradigm science
- Books authored include *Science of New Paradigm* (2014) and *Water of Life: A Cure for Our Body* (2008)
- PhD in Biochemistry, Oxford University, UK (1991)

New Scientific Paradigm

Vacuum is full of particles with negative energy and negative mass, which is an area of information.
Imaginary space forms a 3D field in physical space → Information forms structure.
→ **IMAGINARY SPACE**

Matter consists of physical matter and imaginary matter (*Chi*).
Imaginary matter can be separated from physical matter by physical stimulation.
→ **HOMEOPATHY**

Separated imaginary matter forms a 3D field like that of original matter (3D WAVE).
→ **WATER MEMORY**

Every piece of matter has an imaginary part. Imaginary space does not have time and space.
All pieces of matter are interconnected with one another through imaginary space.
Life and the universe involve a dialogue between the part and the whole.
→ **HOLOGRAM**

A 3D wave can be digitized and expressed as a visual image on a 2D surface.
A digitized 3D WAVE commands imaginary space to form a 3D field of matter around a 2D image.
→ **DIGITAL 3D WAVE**

Figure 1

cannot be compared to life in the spacious, outside world.

Experiments will be easily forgotten if there is no theory explaining the observations. Like quantum science, which was developed in the twentieth century to explain unexplainable facts such as black-body radiation, we need a new paradigm. As shown by Dr. Delorme, there are many facts that cannot be explained by the current materialist paradigm. I have proposed a new scientific paradigm to explain water memory. Homeopathy, which utilizes water memory, has more than 200 years of history, but it cannot be explained by a materialist paradigm.

I do not have time for a detailed explanation. A summary of my new scientific paradigm is given in **Figure 1**. According to Paul Dirac, space is full of particles with negative energy and negative mass. My hypothesis starts from Dirac's sea of vacuum, which is imaginary space. Matter consists of a physical aspect and an imaginary aspect. I named the imaginary aspect of matter as a *3D wave,* but it is similar to the concept of *Chi* in the Orient.

In oriental thinking, every piece of matter has its own *Chi.* Physical matter has a corresponding imaginary 3D wave, expressed as a field with a 3D structure in physical space. 3D fields expressed by imaginary 3D waves can be referred to as *information*. Every biological reaction is an interaction between 3D waves.

A 3D wave can be separated from matter by physical stimulation, such as vigorous shaking or pounding in homeopathy, or electrical stimulation via Schumann waves, which is the resonant frequency of the Earth. The separated 3D wave still functions like the original substance. This is called water memory.

Water memory can also be digitized. The late Dr. Jacques Benveniste used white noise for digitization. I digitized water memory with visual imaging. A digitized 3D wave can be stored in a computer, and it can be expressed in various ways. A digitized 3D wave also functions like the original substance.

A digitized 3D wave has many advantages because it is very stable—it does not lose its functionality as time passes. It has very wide applications. By comparison, analog-type water memory, whether direct or indirect, has very limited applications. But if you can digitize a 3D wave, especially with a visual image, then it can be expressed on a card or in clothes or in wallpaper, so you can easily make a healing environment with printed images. A 2D image of the substance expressed on a flat surface forms a 3D field around a card, as displayed in **Figure 2**.

UN (Healing Energy) Card

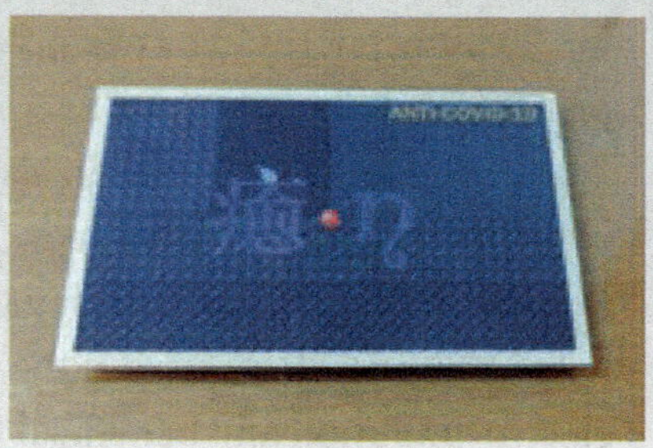

A 3D wave of matter can be digitized and stored in a computer.

UN card commands imaginary space to form a 3D field of matter around the card.

A specific 3D wave of a medically effective substance can be provided with an image (card, wallpaper, cloth, ...)

For example, when P53, a tumor suppressor, is expressed on a UN card, it functions like P53 in a human. Likewise, an anti-COVID-19 UN card protects a person from coronavirus.

Digital 3D waves of more than 1,000 substances are stored in a computer.

Figure 2

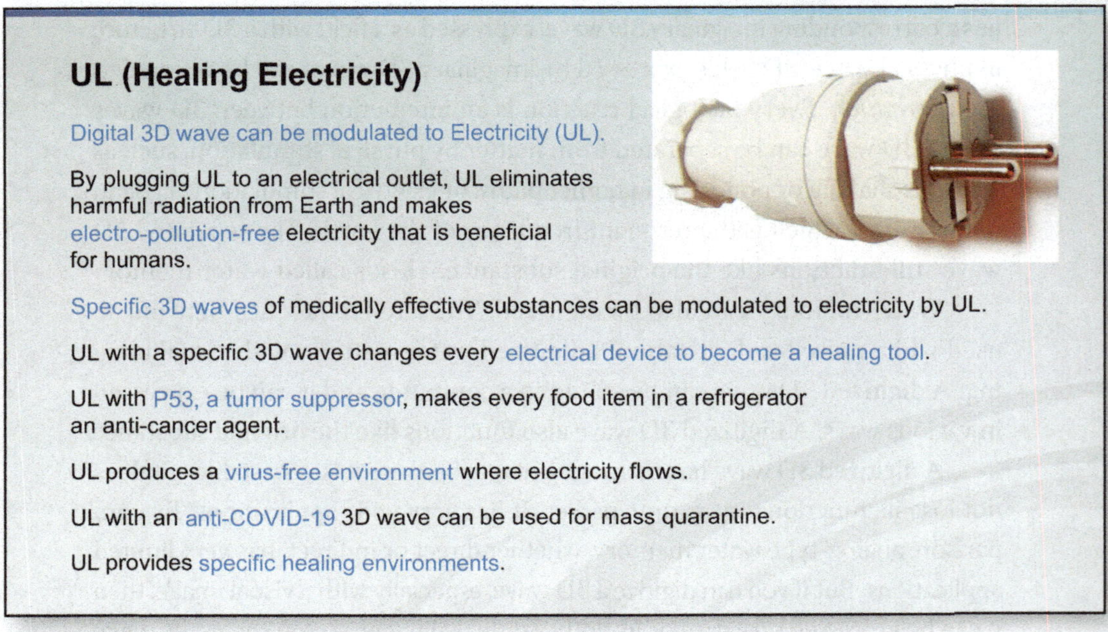

Figure 3

Why is this kind of card, which is printed with a normal printer and normal ink, effective? This card commands the imaginary space to form a 3D field of the substance around the card, hereafter referred to as "UN," a name for "healing energy." For example, a UN card that expresses P53, a tumor suppressor protein, functions as a tumor suppressor.

A digitized 3D wave can also be modulated to electricity using an electrical plug, as shown in **Figure 3**. The device is referred to as "UL," a name for "healing electricity." By plugging the UL into an electrical outlet, you can eliminate harmful radiation from the Earth and purify electricity to make it free of electro-pollution and beneficial for humans. Specific digital 3D waves of medically effective substances can also be put into the UL, so that the space where electricity flows will protect people against specific diseases. A specific healing environment can be provided with a variable electrical code.

The field created by the UN card and the UL plug is in the imaginary area, which does not have time and space, which is a condition for a hologram. The experiments by Dr. Delorme showing that we are not separated from one another and our environment can be explained with this new scientific paradigm. As mentioned above, all matter consists of a physical aspect and an imaginary aspect. Imaginary space is basically a hologram. Thus, we are essentially connected with one another and our environment. Our universe and life within it are a hologram, which involves a dialogue between the part and the whole.

The imaginary field created by the UN card and the UL plug can be used for agriculture, livestock, and fish culture, and to suppress viruses. 3D waves of more than 1,000 substances are stored in my computer. They can be combined to cure specific diseases and for general health, or used for industrial purposes as well. The coronavirus that causes COVID-19 can be suppressed by a specific anti-COVID-19 UN card and an anti-COVID-19 UL plug. The card can be used for individual protection, and the plug can be used for mass quarantine. In this manner, the products of new paradigm science provide safe and effective alternatives to vaccines.

The twenty-first century needs a new scientific paradigm beyond materialism, which will expand the horizons of science and benefit human society. ◊

Session 2 | Presentation Two

Biohybridization of Living Organisms: Challenges and Promises

Dr. Serge Kernbach
Director, Research Center of Advanced Robotics and Environmental Science, CYBRES GmbH, Germany

Biohybridization represents a new trend related to the integration of technological systems and living organisms. It started in areas of wearable computing, smart communication, robotic and neural implants, and it is heavily influenced by modern artificial intelligence (AI) and quantum research. Multiple challenges are related to cellular interfaces with electronic systems, energy supply from molecular biochemistry, and biocompatibility of materials, among others. Biohybrid technologies open new horizons for medical applications in terms of prostheses and implants, diagnostics and treatments. For gerontology, they reveal possibilities of digital,

About the Speaker

- Director, CYBRES GmbH, Research Center of Advanced Robotics and Environmental Science, University of Stuttgart, Germany
- Coordinator of several European research projects in the field of biohybrid systems and collective robotics (since 2004)
- Foreign Member of the Russian Academy of Natural Sciences (elected in 2016)
- Main research interest is focused on biological and technological collective systems, robotics, electrochemistry, weak biophysical interactions
- Author or coauthor of 4 books and over 180 articles in international journals and conferences with best paper awards
- Dr. rer. Nat., Electronic Engineering, Robotics

biological, or quantum immortality. For biological and environmental applications, biohybridization opens vistas for biosensors or smart ecosystems.

Research on biohybrid systems reveals two important topics: 1) ultraweak biophysical and biochemical effects that appear on ionic levels and influence the whole organism, and; 2) quantum phenomena in macroscopic systems that shape the cognitive and computational capabilities of human brain. Finally, a combination of these and other research topics allows for the revision offering several unconventional hypotheses and their role in understanding the phenomenon of life.

A New Humanity?

Biohybridization, in simplified form, represents a combination of modern technologies and several scientific research areas that view biosystems in the context of specific biochemical machinery. From this perspective, diseases in biosystems can be repaired and deficiencies can be improved or generally enhanced. Biohybridization aims at a wide range of problems in health care, medicine, ecology, and smart devices, including in surgery, prosthetics, gerontology, the expansion of body functions, and the creation of interfaces to ecosystems. According to various estimates, the market for biohybrid systems will be the largest market in history.

Ideas of combining artificial technology and natural biology are not new. From Frankenstein's creature to synthetic biology, from the Prague Golem to neuroprostheses and cyborgs, such thoughts have occupied human minds for centuries. In fact, even biological evolution can be seen as a very advanced technology developed over millions of years. However, the scientific progress made by humankind introduces a completely new understanding of technology as a new evolutionary force that can transform men and women themselves.

Cultures, faced with the phenomenon of life, have always tried to distinguish it from other phenomena. Life, in the minds of thinkers, did not end but only passed from one form to another, from biological to nonbiological. Later, with the development of modern science in the seventeenth century, the idea arose that biological life is the only possible one. It dominated for nearly 300 years and shaped the scientific philosophy of positivism. Biohybridization revives the idea that nonbiological and mixed forms of conscious life are possible, a notion that supports people's striving for nonbiological immortality.

In this context, technological progress acts as a new creator that engineers a new humanity, based on how scientists understand nature and its processes. As we are now "transforming dreams into reality," developers and designers need to think about what form of new reality we are creating. Do we really understand the nature of reality, taking into account, for example, quantum phenomena? Is life determined only by biochemical processes? Does the new evolution also require a new humanism? Should immortality be developed only for the richest people (new gods)? to everyone? is everyone already immortal? This discussion looks at some of these technologies and reinterprets the old questions of life.

Biohybridization Scenarios: Synthetic Biology and Advanced Neurorobotic Implants

Biohybridization, as a combination of technologies to improve the body, is a dynamically developing scientific and technological area. At the moment, two clusters of technologies are actively growing: 1) one with a focus on synthetic biology and 2) one with a predominance of elements from robotics, neuro- and materials science, and biointerfaces.

In the first case, synthetic biology targets programmable biosystems, whose development is significantly ahead of the information technology (IT) sector. Startups in synthetic biology in 2019 raised $1.9 billion, and in 2020 $3 billion. They cover projects from the creation of enzymes and alternative meats to microbes and more complex life forms. Interesting areas are tissue engineering, the potential goal of which is to create biological organs for transplantation, and the method of gradual DNA editing (genome editing was recognized as the method of the year in 2011) (Stevens 2017). In 2017, the genome was first edited in the human body, and in 2019 the world's first genetically modified person was born.

The second case targets the hybridization of existing organisms, a process that already began about 15–20 years ago with the development of robotic prostheses and wearables. The new concept of the neuromuscular-skeletal prosthesis represents an autonomous robot directly connected to the nerves, muscles, and skeleton of a person. Control signals are read using electrodes implanted in nerves and muscles. Prostheses send tactile sensory feedback to the nervous system to determine how hard to grasp or squeeze an object. The prosthesis provides a sensation that the user perceives as coming from a missing limb (**Figure 1**). The key technologies for such prostheses are neurointerfaces that allow reading signals from neurons and stimulating them. At the moment, several independent technological areas have already been developed, such as brain-computer interfaces (for example, for games), techniques for implanting electrodes into the brain (see, for instance, Neuralink), functional brain mapping, neuroengineering, and so forth.

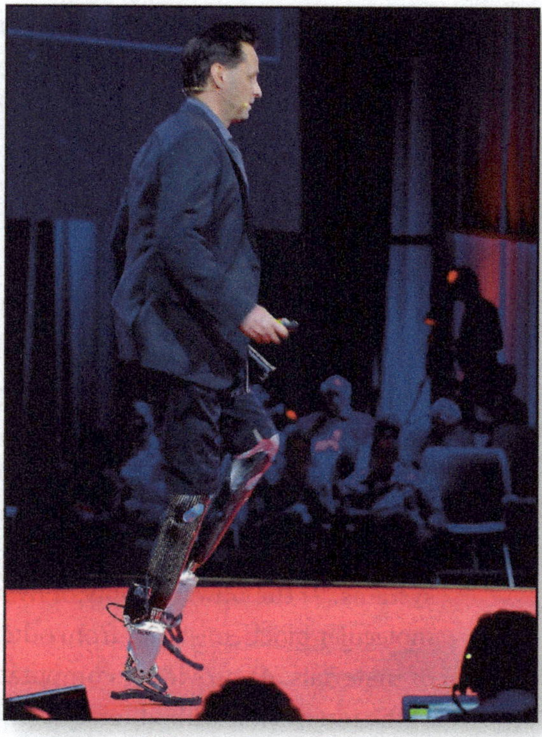

Figure 1 Hugh Herr demonstrating new robotic prosthetic legs at TED Talk 2014. That was the first demonstration of a running gait under neural command. "The more I fire my muscles, the more torque I get," Herr said. (Source: https://commons.wikimedia.org/wiki/File:Hugh_Herr,_TED_2014.jpg)

The final component of biohybridization is the fusion of cellular and noncellular components, such as the bioartificial liver project. One of the successful areas here is 3D bioprinting technology, which allows for the creation of soft tissues, bones, blood vessels, and more-complex organs for surgery. For example, in 2017, a patient received an artificial ear created using bioprinting technology (Zhou et al. 2018).

With the development of wearable computers and communications, significant computing power has become available to biomicrosystems. This allows integrating AI with biocomponents. A relatively new area is the development of quantum biology, which radically revises the physical and chemical foundations of biosystems.

Biohybrid Interfaces and Related Ecologies

Each path of biohybridization has one or another advantage that the associated systems and technologies provide. Solving technological problems at one level immediately opens up new opportunities at another level. For example, improving the method of implanting electrodes into the brain makes it possible to treat chronic depression, which changes behavior and personality, thus affecting consciousness. In general, biohybridization is characterized by several fundamental areas that are traditionally associated with biohybrid interfaces: 1) the interaction of cells and organs with electronic and mechanical systems, 2) the power supply of hybrid components (whether derived from molecular biochemistry or from other energy sources), 3) the biocompatibility of materials, 4) providing computational power and communications, and 5) the interaction with other biocomponents, organisms, and biosystems.

Biohybridization occurs not only with animals but with plant tissues and organisms. All elements of biohybrid interfaces arise here: from energy supply and biocompatibility to interaction with other systems. Since ethical and technological aspects are much simpler with plant organisms, researchers have undertaken many projects with biohybrid plants, including

- The Flora Robotica project, which combines plants and urban architecture, creating "smart" plants (Hamann et al. 2015).
- The watchPlant project (García-Carmona et al. 2021) (**Figure 2**), linking urban plants (trees) into an information sensor network. This project is designed to get energy from the biochemistry of plant sap, allowing plants to sense air quality and chemical and electromagnetic pollution by plants. An interesting aspect of this project is a computer interface between natural urban and forest ecologies, and urban infrastructure.

An artist's image of a biohybrid plant with multiple physiological sensors from the watchPlant project, funded by the European Union. (Source: Serge Kernbach) **Figure 2**

- The Biohybrids project, which is developing an ionic interface between cells and electronics (Kernbach et al. 2022) and provides access to quantum level phenomena, such as nuclear spins and ortho-/para-isomers of water (Kernbach 2022).

Biohybridization and Rethinking the Phenomenon of Life I: Nonchemical Biological Mechanisms

In the study of biohybrid systems and interfaces, researchers face many effects whose mechanisms are not yet fully understood. First, there are weak and ultra-weak signals that are of a clear nonchemical nature. These studies began in the 1930s and are associated with the names of Gurvich, Magre, Kaznacheev, Popp, and others. Distant nonchemical interactions arise during optical and electromagnetic contact between cell cultures. They affect the rate of cell growth and protein content, the appearance of a cytopathic effect (upon optical contact with pathogenic cultures), and abnormal cell development. Popp and coauthors focused on the quantum properties of photons; in general, this hypothesis about ultra-weak photon emission as a messenger for information transfer in cell-to-cell communication has received recent confirmations (Prasad et al. 2014). Maslobrod's experiments show the possibility of macroquantum entanglement between germinating triticale grains, when chromosomal dysfunctions in one part of the grains are transmitted to other, distant parts (Maslobrod 2012).

Similar mechanisms are demonstrated in complex biochemical reactions. For example, the polymerase chain reaction is capable of noncontact replication of remote parts of DNA in a weak magnetic field, as in Montagnier's experiment (Montagnier et al. 2015). These works are similar to quantum teleportation, which is used in the quantum Internet, where, for example, the possibility of teleportation of qubits into diamond crystals is demonstrated (Tsurumoto et al. 2019). Since such distant interactions change the ionic content of cellular and aqueous media—for example, Ca^{2+}, OH^-, and H^+ ions—they provide tools based on electrochemical methods for studying these processes.

Biohybridization and Rethinking of the Phenomenon of Life II: Achieving Potential Immortality

When connecting neuroimplants and controlling roboprostheses, it becomes possible to digitize cognitive functions of the brain and transfer them to a computer simulation. One of the first projects in this area investigated simple model organisms such as the *Caenorhabditis elegans* worm (Sarma et al. 2018). About a thousand cells of this organism are implemented in a computer model (the OpenWorm project); 302 nerve cells and 95 muscle cells allow the worm to move in the virtual world of a computer simulation.

Microsoft's Cyber All Project, or Eterni.me, is considering a similar idea: The preservation of personal information allows creating a digital imprint of an individual and its possible manifestation in a digital world. There are even registered patents for the technology of creating chatbots based on the personal data of deceased people.

Biohybridization provides another way to significantly prolong active life and to approach the potential of biological immortality by stabilizing body functions and replacing worn-out organs. Bank of America has estimated the life-extension market at $600 billion by 2025. There are already dozens of firms operating here, focusing on everything from genomics and AI applications to smart health devices and biohacking. Almost all IT giants, such as Google, Amazon, and Apple, are investing in firms devoted to this area.

There is yet another hypothesis, which can be described as *quantum immortality.* One of its interpretations was given by Roger Penrose and Stuart Hameroff, and new experimental data has appeared in favor of this model (Jedlicka 2017). The basic idea is that qubits can be generated in biological neurons. In the process of neuronal calculations (information processing in biological way) micro and macro levels occurs, and a quantum computing system is formed. Such a quantum computer always accompanies biological consciousness. Unusual abilities of consciousness are frequently associated with such a quantum machinery. This hypothesis assumes that the death of biological consciousness releases a 'nonbiological consciousness' that persists for at least some time. There are many projects and publications that explore old traditions and techniques from Tibet, China, Japan, and India dealing with achieving potential immortality of such 'nonbiological consciousness' (Kozhevnikov et al. 2013).

Conclusion

This short overview highlights some perspectives and open questions in a new technological area related to the biohybridization of living organisms. The combination of different technological solutions, such as cell implants, robotic prostheses, and semiconductors for human modification, has the potential to solve human health problems and increase the quality and duration of active life. Moreover, the integration of technologies into biosystems provides better understanding of the nonchemical mechanisms of cellular regulation associated with weak and ultraweak signals and long-range interactions. At the ecological level, biohybridization of plants and microorganisms makes it possible to effectively balance ecosystems; to establish networks of forest, urban, and animal systems; and to develop new living biosensors.

In addition to opening up a huge market and a having significant impact on social development, this technological area changes the concept of evolution, raising serious arguments for achieving potential immortality. Experiments with biointerfaces based on ionic electrochemical systems allows for the exploration of nonchemical mechanisms that supposedly have the nature of quantum effects in macrosystems. These mechanisms can be responsible for a number of unusual phenomena attributed to consciousness and can ultimately allow better understanding of the phenomenon of conscious life. ◇

References

García-Carmona, L., S. Bogdan, A. Diaz-Espejo, M. Dobielewski, H. Hamann, V. Hernández-Santana, A. Kernbach, S. Kernbach, A. Quijano-López, A. Quijano-López, N. Roxhed, B. Salamat, and M. Wahby. 2021. "Biohybrid systems for environmental intelligence on living plants: WatchPlant project." GoodIT '21: *Proceedings of the Conference on Information Technology for Social Good:* 210-215. https://dl.acm.org/doi/10.1145/3462203.3475885

Hamann, H., M. Wahby, T. Schmickl, P. Zahadat, D. Hofstadler, K. Stoy, S. Risi, A. Faina, F. Veenstra, S. Kernbach, I. Kuksin, O. Kernbach, P. Ayres, and P. Wojtaszek. 2015. "Flora Robotica – Mixed Societies of Symbiotic Robot-Plant Bio-Hybrids." *2015 IEEE Symposium Series on Computational Intelligence:* 1102–1109. https://doi.org/10.1109/SSCI.2015.158.

Jedlicka, P. 2017. "Revisiting the Quantum Brain Hypothesis: Toward Quantum (Neuro)biology?" *Frontiers in Molecular Neuroscience* 10: 366. https://doi.org/10.3389/fnmol.2017.00366.

Kernbach, S. 2022. "Electrochemical Characterization of Ionic Dynamics Resulting from Spin Conversion of Water Isomers." *Journal of The Electrochemical Society* 169: 067504. https://doi.org/10.1149/1945-7111/ac6f8a

———, O. Kernbach, I. Kuksin, A. Kernbach, Y. Nepomnyashchiy, T. Dochow, and A. Bobrov. 2022. "The biosensor based on electrochemical dynamics of fermentation in yeast *Saccharomyces cerevisiae.*" *Environmental Research* 213: 113535. https://doi.org/10.1016/j.envres.2022.113535

Kozhevnikov, M., J. Elliott, J. Shephard, and K. Gramann. 2013. "Neurocognitive and Somatic Components of Temperature Increases during g-Tummo Meditation: Legend and Reality." *PLOS ONE* 8 (3). https://doi.org/10.1371/journal.pone.0058244.

Maslobrod, S. 2012. "Effect of a Long-Range Interaction Appeared between Germinating Seeds." *Electronic Processing of Materials* 48 (6): 99–113.

Montagnier, L., E. Del Giudice, J. Aïssa, C. Lavallee, S. Motschwiller, A. Capolupo, A. Polcari, P. Romano, A. Tedeschi, and G. Vitiello. 2015. "Transduction of DNA Information through Water and Electromagnetic Waves." *Electromagnetic Biology and Medicine* 34 (2): 106–112. https://doi.org/10.3109/15368378.2015.1036072.

Prasad, A., C. Rossi, S. Lamponi, P. Pospíšil, and A. Foletti. 2014. "New perspective in cell communication: Potential role of ultra-weak photon emission." *Journal of Photochemistry and Photobiology B: Biology* 139: 47–53. https://doi.org/10.1016/j.jphotobiol.2014.03.004

Sarma, G.P., C.W. Lee, T. Portegys, V. Ghayoomie, T. Jacobs, B. Alicea, M. Cantarelli, M. Currie, R.C. Gerkin, S. Gingell, P. Gleeson, R. Gordon, R.M. Hasani, G. Idili, S. Khayrulin, D. Lung, A. Palyanov, M. Watts, and S.D. Larson. 2018. "OpenWorm: overview and recent advances in integrative biological simulation of *Caenorhabditis elegans*." *Philosophical Transactions of the Royal Society B: Biological Sciences* 373 (1758). https://doi.org/10.1098/rstb.2017.0382.

Stevens, S. 2017. "Synthetic Biology in Cell and Organ Transplantation." *Cold Spring Harbor Perspectives in Biology* 9 (2): a029561. https://doi.org/10.1101/cshperspect.a029561.

Tsurumoto, K., R. Kuroiwa, H. Kano, Y. Sekiguchi, and H. Kosaka. 2019. "Quantum teleportation-based state transfer of photon polarization into a carbon spin in diamond." *Communications Physics* 2 (74). https://doi.org/10.1038/s42005-019-0158-0.

Zhou, G., H. Jiang, Z. Yin, Y. Liu, Q. Zhang, C. Zhang, B. Pan, J. Zhou, X. Zhou, H. Sun, D. Li, A. He, Z. Zhang, W. Zhang, W. Liu, and Y. Cao. 2018. "*In Vitro* Regeneration of Patient-specific Ear-shaped Cartilage and Its First Clinical Application for Auricular Reconstruction." *EBioMedicine* 28: 287–302. https://doi.org/10.1016/j.ebiom.2018.01.011.

Session 2 | First Commentary
On the Presentation by Dr. Serge Kernbach

Concerning Health Benefits from New Paradigm of Science

Professor Roumiana Tsenkova
Aquaphotomics Research Department,
Kobe University, Japan

In alignment with the theme of the current session, "The Promise of New Paradigm Science in Solving Environmental Problems," Dr. Serge Kernbach gave a presentation on the topic, "Health Benefits from New Paradigm Science." The speech covered the new trend of biohybridization, an integration of technological systems and living organisms, aimed at solving a wide range of problems that prevent a higher quality of life in general. This approach, in short, views biosystems as "machines" or "mechanisms" that can be repaired, improved, or enhanced.

Judging by the popularity of superheroes, cyborgs, people with extra

About the Speaker

- Founder and Professor, Aquaphotomics Research Department, Kobe University, Japan
- Primary area of interest is the use of near-infrared spectroscopy (NIRS) and multivariate analyses for biodiagnostics and biomonitoring related to functional studies in life science, biotechnology, and agriculture
- She was the first to apply NIRS for disease diagnostics
- Proposed, in 2005, a new discipline named Aquaphotomics, to build up a database of water absorbance bands specific for each biological and aqueous system called an aquaphotome, and to use it for diagnosis and characterization on a system
- Dr. Eng. (Moscow Academy of Agricultural Sciences and Bulgarian Russe Univ.), Dr. Agr. (Hokkaido Univ.)

abilities, robots, spaceships, virtual reality, and similar topics in the entertainment industry, one can clearly observe that contemporary society is quite fascinated with the vision of a future manmade world. I deliberately use the term "manmade" because, while people certainly excel in the production of machines and mechanisms, they still have neither reached the understanding of what "life" is, nor can they create a living being from scratch.

The "mechanism" of life and, for example, what consciousness is, are beyond current understanding. Dr. Kernbach thus rightly posits some age-old questions, such as: What is life? What is the nature of reality? Do we have enough of an understanding of the systems we are trying to repair or improve?

As a founder of aquaphotomics (Tsenkova 2009), a new scientific discipline studying aqueous and biological systems through the prism of their interaction with electromagnetic radiation thanks mainly to the water molecular system, I can add to the complexity of the presented topic with additional insights from this complementary field. The purpose is the same—improved quality of life on all levels. Aquaphotomics was established in 2005, with more and more publications and contributions from different fields coming from more than 15 laboratories around the world (Tsenkova 2009; Munćan and Tsenkova 2019).

The current, mainly reductionistic, approach in science needs a perspective of the wholeness of systems and causality. Therefore, methods for noninvasive, real-time monitoring of systems' functionality are crucially required (Tsenkova 2009). Furthermore, the role of the water molecular system in all these processes and the relation between water structure and the resultant functionality of these systems need to be explored.

Aquaphotomics has been established as a complementary discipline to current "omics" sciences (**Figure 1**), which are all reductionistic, following single molecules or, recently, organ components, such as DNA and proteins, while aquaphotomics follows water as a complex molecular system (Munćan and Tsenkova 2019).

One of the main tools for representation and rapid insight into the connection of water structure with the functionality of the respective system is

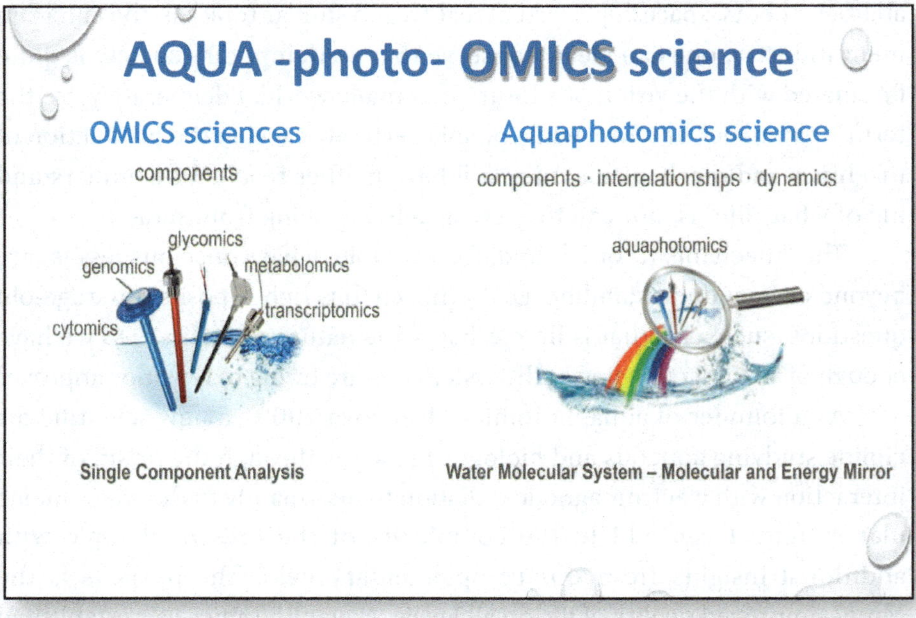

Figure 1 Credit: All figures courtesy of Prof. Roumiana Tsenkova

the aquagram (**Figure 2**), a graph that displays the light absorbance spectral pattern of water (Tsenkova 2010; Tsenkova et al. 2018). For the near-infrared first overtone spectral region of water, there are 12 specific absorbance bands discovered that originate from different water molecular conformations (Tsenkova 2009; Kuroki et al. 2019).

The next step of aquaphotomics is organizing a database of water spectral patterns, or aquagrams, into what is called aquaphotomes (Tsenkova 2009). An aquaphotome is an ever-growing database, where aquagrams of different aqueous and biological systems, under different perturbations and different conditions, are stored, matching the water spectral pattern of the system with its respective functional state. While aquaphotomics still does not have the answer to what life is, having an aquaphotome reveals to us the connections between seemingly unrelated phenomena involving water and thus brings a better understanding of the steps of this "life mechanism." For example, we discovered similarity in the aquagrams of fullerol in water solu-

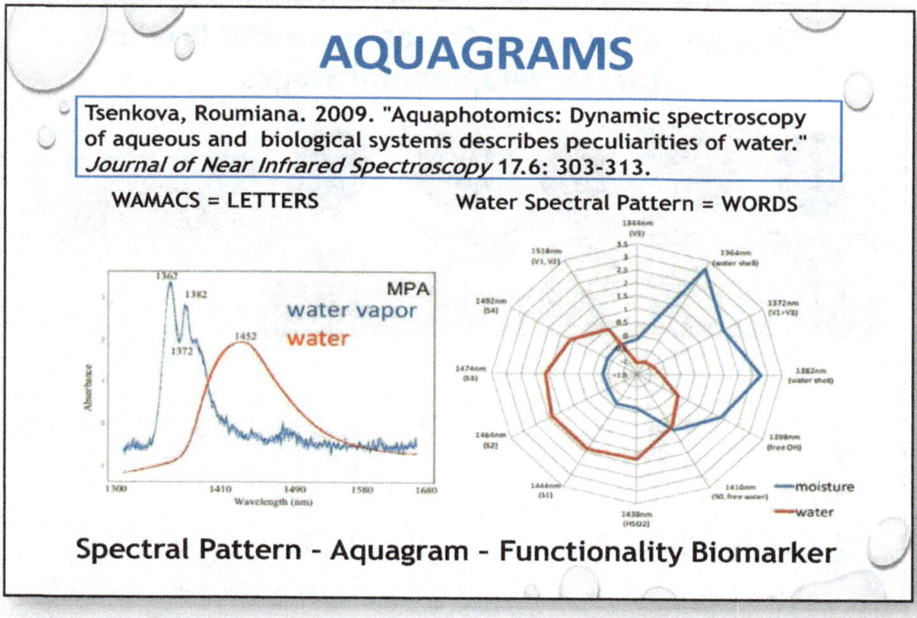

Credit: Roumiana Tsenkova

Figure 2

tions in different (small) concentrations (Matija et al. 2012) and aquagrams of the hydrophilic ionomer nafion (Munćan et al. 2016), immersed in water and subjected to consecutive irradiations.

It has also been revealed that the common water molecular structures in various cases—namely, hydrogen-bonded water, confined water (trapped water), and protonated water clusters—are the main water structures in cells as well. The influence of the concentration of nanomaterial and the influence of light resulted in a similar reorganization of the water molecular structure in those systems. There are many implications of this finding, not only in the field of nanomaterials, self-organization, and interface science, but also in the significance of light for life.

In another case, for example, using noninvasive monitoring of larvae development (Tsugehara, Tsenkova, and Takeda 2006), the absorbance spectral pattern of water in this organism could allow prediction of its developmental phase and whether it will develop or not, showing that

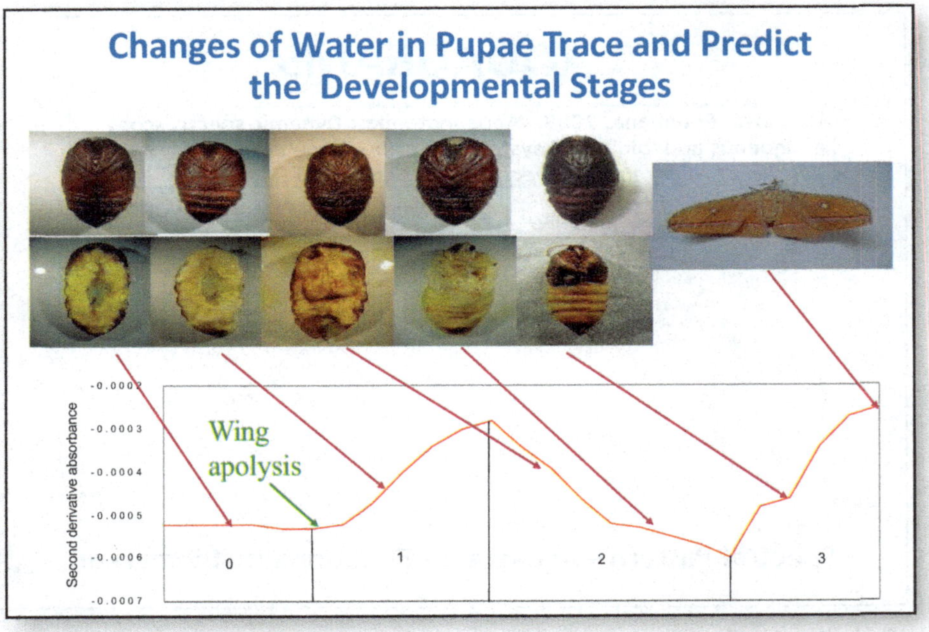

Figure 3 Credit: Roumiana Tsenkova

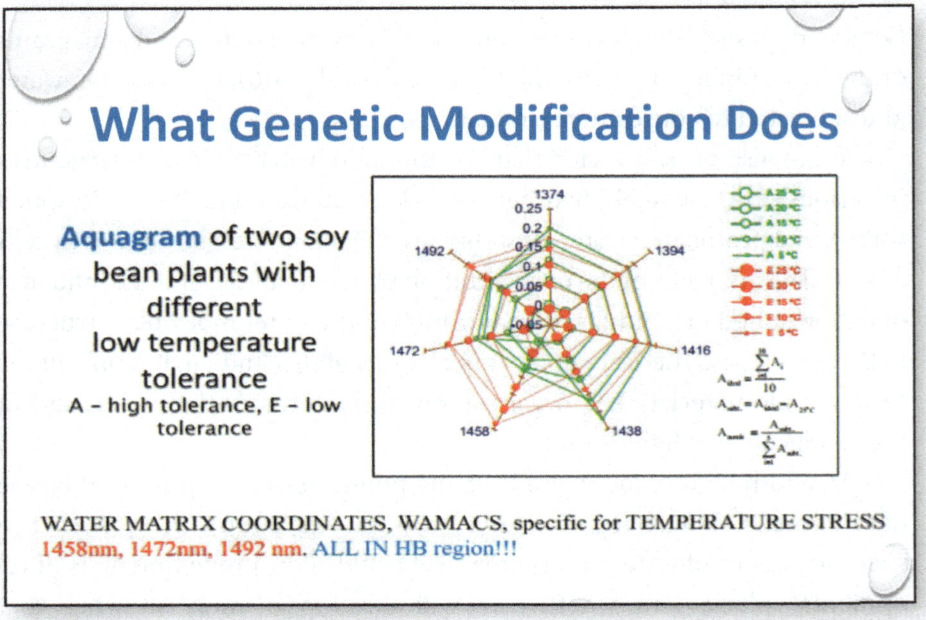

Figure 4 Credit: Roumiana Tsenkova

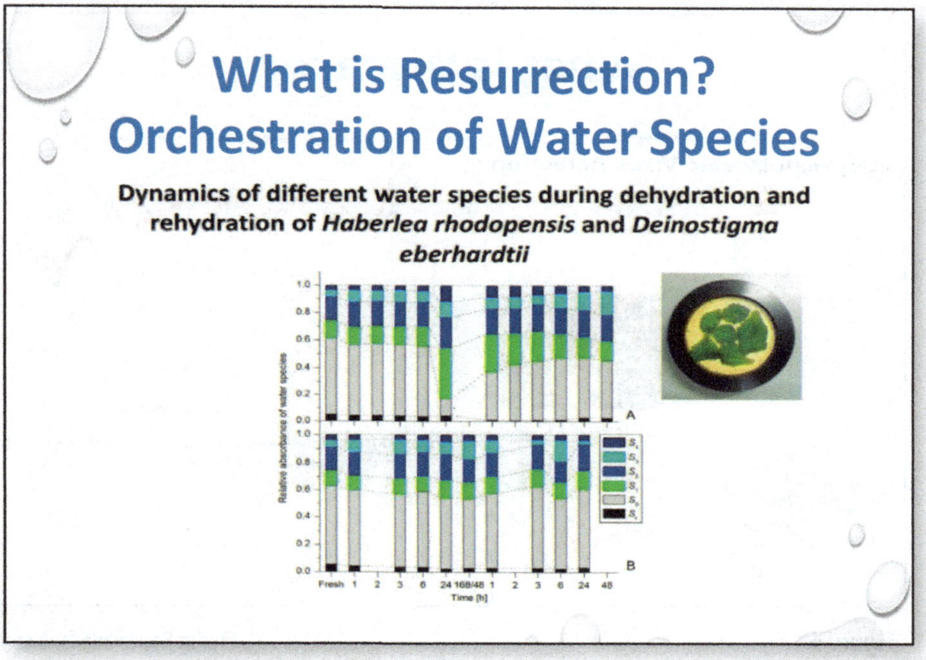

Credit: Roumiana Tsenkova **Figure 5**

information on life, stages of life or death can be read and even predicted from its water spectrum time series (**Figure 3**).

On another issue, using aquagrams we were able to determine what genetic modification actually does. Comparison of aquagrams of leaves of different soybean cultivars, genetically modified to have a different tolerance to cold (Jinendra 2011), showed that the cold-tolerance trait is actually an ability of the plant to keep the water structure in leaves in a weakly hydrogen-bonded state, despite the decrease in temperature (**Figure 4**). This finding showed that the final result of genetic modification is the production of substances in leaves, which act as heaters or thermostats, keeping the water structure in the desired state (Munćan, Aouadi, and Tsenkova 2020).

A similar mechanism of "orchestrating the water structure" (**Figure 5**) was observed in a study of desiccation tolerance in resurrection plants (Kuroki et al. 2019). The plants that can survive extremely long periods with-

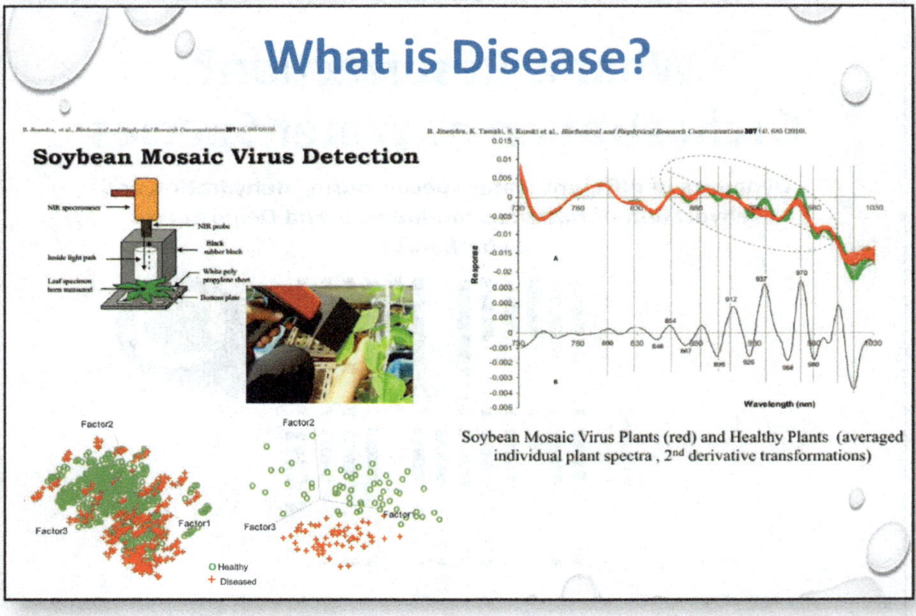

Figure 6 Credit: Roumiana Tsenkova

out water have the ability to synchronize the changes of all water species (Kuroki et al. 2019; Munćan, Aouadi, and Tsenkova 2020). These studies provided new insights into the stress response in plant tissues, and the results may relate to their longevity as well. Similarly, the viral infection of plants was shown to be caused by the loss of certain water molecular structures (Jinendra et al. 2010; Jinendra 2011) (**Figure 6**).

Astonishingly, similar observations were found in higher-level organisms and diseases such as prion protein disease (Tsenkova 2005; Tsenkova et al. 2014), mastitis (Tsenkova et al. 2000, 2001, 2006) and others (Sakudo et al. 2004; Kinoshita et al. 2012; Chatani et al. 2014; Munćan and Tsenkova 2019). The study of prion protein disease in mouse models led to another surprising finding: the spectral patterns of the mouse abdomen provided earlier detection of prion disease when compared to the head, which shines light on the origin and the pathways of the disease (**Figure 7**).

These few examples briefly illustrate the connectedness between many

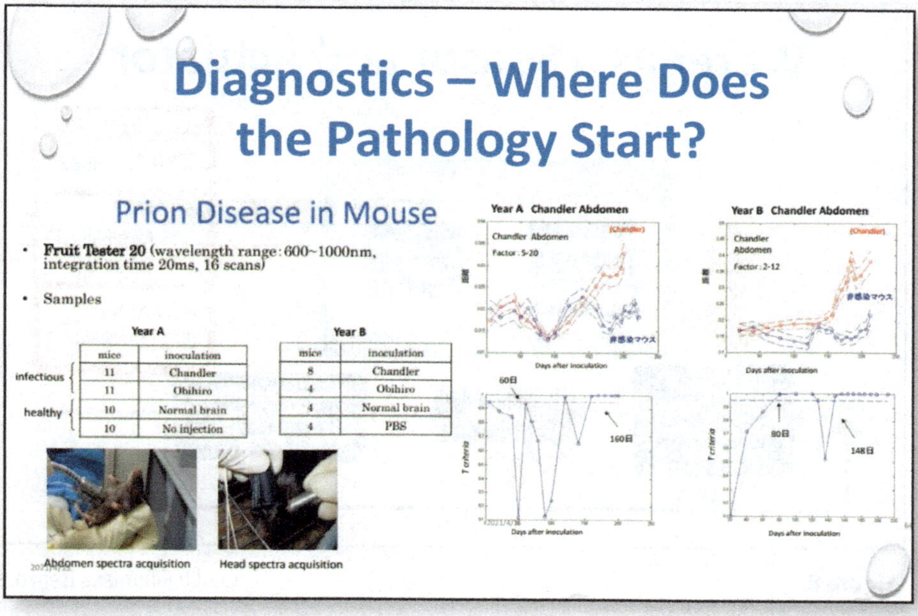

Credit: Roumiana Tsenkova **Figure 7**

phenomena in living organisms, caused by the existence of a common platform, namely, the interconnected water molecular species, and especially between them and the water molecular structure and its functionality. While this better understanding per se is a great step towards better bio-design, the most fascinating fact is that it is obtained through water. What enabled this progress is the discovery that, under various perturbations, water can be a sensor and an actuator (Tsenkova 2009; Munćan and Tsenkova 2019; Van de Kraats, Munćan, and Tsenkova 2019) (**Figure 8**). The interaction of water with light at all frequencies, because water absorbs light of the entire EM spectrum, provides us with a magnificent tool for noninvasive observation and control of living processes in real time. Water–light interactions provide immense information for water design and engineering toward system biomodulation and repair, enhancement, disease prevention, and healing of organisms at the level where all such processes begin using only natural building blocks and energies. ◊

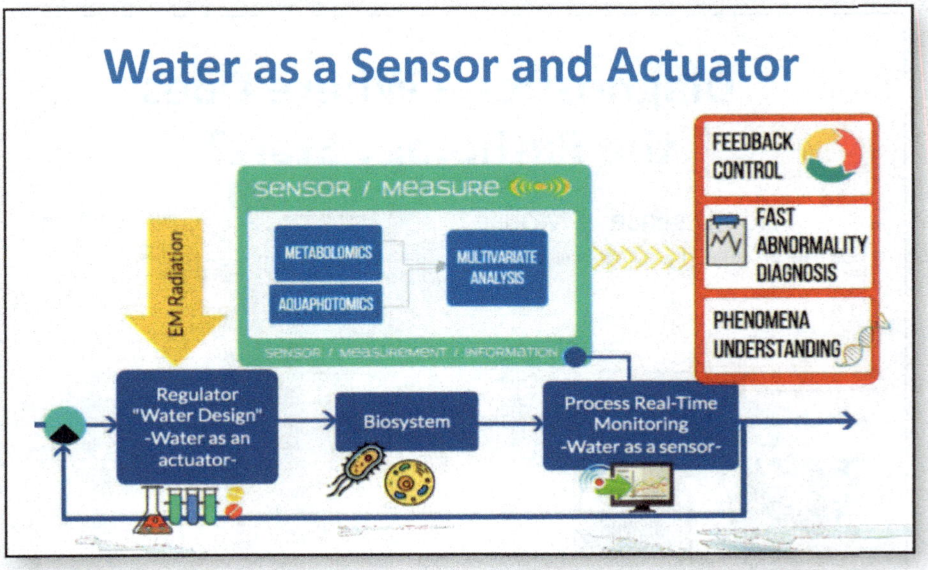

Figure 8 Credit: Roumiana Tsenkova

References

Chatani, E., Y. Tsuchisaka, Y. Masuda, and R. Tsenkova. 2014. "Water Molecular System Dynamics Associated with Amyloidogenic Nucleation as Revealed by Real Time Near Infrared Spectroscopy and Aquaphotomics." *PLOS ONE* 9 (7): e101997. https://doi.org/10.1371/journal.pone.0101997.

Jinendra, B. "Near Infrared Spectroscopy and Aquaphotomics: Novel Tool for Biotic and Abiotic Stress Diagnosis of Soybean." (PhD dissertation, Kobe University, 2011).

———, K. Tamaki, S. Kuroki, M. Vassileva, S. Yoshida, and R. Tsenkova. 2010. "Near Infrared Spectroscopy and Aquaphotomics: Novel Approach for Rapid In-vivo Diagnosis of Virus Infected Soybean." *Journal of Biochemical and Biophysical Research Communications* 397 (4): 685–690. https://doi.org/10.1016/j.bbrc.2010.06.007.

Kinoshita, K., M. Miyazaki, H. Morita, M. Vassileva, C. Tang, D. Li, O. Ishikawa, H. Kusunoki, and R. Tsenkova. 2012. "Spectral pattern of urinary water as a biomarker of estrus in the giant panda." *Scientific Reports* 2: 856. https://doi.org/10.1038/srep00856

Kuroki, S., R. Tsenkova, D. Moyakova, J. Munćan, H. Morita, S. Atanassova, and D. Djilianov. 2019. "Water molecular structure underpins extreme desiccation tolerance of the resurrection plant Haberlea rhodopensis." *Scientific Reports* 9: 3049. https://doi.org/10.1038/s41598-019-39443-4.

Matija, L., R. Tsenkova, M. Miyazaki, K. Banba, and J. Munćan. 2012. "Aquagrams: Water Spectral Pattern as Characterization of Hydrogenated Nanomaterial." *FME Transactions* 40 (2): 51–56.

Munćan, J., and R. Tsenkova. 2019. "Aquaphotomics—From Innovative Knowledge to Integrative Platform in Science and Technology." *Molecules* 24 (15): 2742. https://doi.org/10.3390/molecules24152742

———, B. Aouadi, and R. Tsenkova. 2020. "New perspectives in plant and plant-based food quality determination: Aquaphotomics." *Plant-based Bioactives, Functional Foods, Beverages and Medicines: Processing, Analysis and Health Benefit.* Edited by M.R. Goyal, Z. Kovacs, A. Nath, and H. Suleria. (Palm Bay, FL: Apple Academic Press, Inc. under Taylor and Francis).

———, S. Kuroki, D. Moyankova, H. Morita, S. Atanassova, D. Djilianov, and R. Tsenkova. 2019. "Recent advancements in plant aquaphotomics – Towards understanding of 'drying without dying' phenomenon and its implications." *NIR news* 30 (5-6): 22–25.

———, I. Mileusnic, V. Matovic, J.S. Rosic, and L. Matija. 2016. "The prospects of aquaphotomics in biomedical science and engineering." Aquaphotomics: Understanding Water in Biology – 2nd International Symposium, Kobe University, Kobe, Japan. December 26–29, 2016.

Sakudo, A., R. Tsenkova, K. Ikuta, and T. Onodera. 2004. "NIRS for diagnosis of HIV and other viruses and prion disease." Japanese patent application No 2004-329249.

Tsenkova, R. 2005. "Visible-near infrared perturbation spectroscopy: Water in action seen as a source of information," in NIR in action—making a difference. *Proceedings of the 12th International Conference on Near-infrared Spectroscopy.* Edited by G. Burling-Claridge, S. Holroyd, and R. Sumner. (Hamilton, New Zealand: The New Zealand Near Infrared Spectroscopy Society Inc.), 607–612.

———, 2009. "Aquaphotomics: Dynamic Spectroscopy of Aqueous and Biological Systems Describes Peculiarities of Water." *Journal of Near Infrared Spectroscopy* 17: 303–314. https://doi.org/10.1255/jnirs.869.

———, 2010. "Aquaphotomics: Water in the biological and aqueous world scrutinised with invisible light." *Spectroscopy Europe* 22 (6): 6–10. https://www.spectroscopyeurope.com/article/aquaphotomics-water-biological-and-aqueous-world-scrutinised-invisible-light. Accessed on May 19, 2022.

———, S. Atanassova, S. Kawano, and K. Toyoda. 2001. "Somatic cell count determination in cow's milk by near-infrared spectroscopy: a new diagnostic tool." *Journal of Animal Science* 79 (10): 2550–2557. https://doi.org/10.2527/2001.79102550x.

———, S. Atanassova, H. Morita, K. Ikuta, K. Toyoda, I. Iordanova, E. Hakogi. 2006. "Near Infrared Spectra of Cows' Milk for Milk Quality Evaluation: Disease Diagnosis and Pathogen Identification." *Journal of Near Infrared Spectroscopy* 14 (6): 363–370. https://doi.org/10.1255/jnirs.661.

———, I. Iordanova, K. Toyoda, and D.R. Brown. 2004. "Prion protein fate governed by metal binding." *Biochemical and Biophysical Research Communications* 325 (2): 1005–1012. https://doi.org/10.1016/j.bbrc.2004.10.135.

———, J. Munćan, B. Pollner, and Z. Kovacs. 2018. "Essentials of Aquaphotomics and Its Chemometrics Approaches." *Frontiers in Chemistry* 6: 363. https://doi.org/10.3389/fchem.2018.00363.

———, K. Murayama, S. Kawano, Y. Wu, K. Toyoda, and Y. Ozaki. 2000. "Near infrared spectroscopy for mastitis diagnosis: Two-dimensional correlation study in short wavelength region." *Two-Dimensional Correlation Spectroscopy.* Edited by Y. Ozaki and I. Noda. American Institute of Physics Conference Proceedings 503: 307–311. https://doi.org/10.1063/1.1302884.

Tsugehara, T., R. Tsenkova, and M. Takeda. 2006. "A non-invasive monitoring of developmental changes in pupal case of Anthereaea pernyi by near-infrared spectroscopy." *International Journal of Wild Silkmoth & Silk* 11: 14–20.

Van de Kraats, E.B., J. Munćan, and R. Tsenkova. 2019. "Aquaphotomics – Origin, concept, applications and future perspectives." *Substantia* 3 (2): 13–28. https://doi.org/10.13128/Substantia-702.

Session 2 | Second Commentary
On the Presentation by Dr. Serge Kernbach

Regarding Biohybridization of Living Organisms: Challenges and Promises

Dr. Konstantin Korotkov
President, International Union for Medical and Applied Bioelectrography, Israel

Dr. Serge Kernbach has presented a new trend in the development of electronics capable of interfacing with the nervous system. It is a rapidly advancing field, with applications in basic science and clinical translation. Devices containing arrays of electrodes can be used in the study of cells grown in culture, or they can be implanted into damaged or dysfunctional tissue to restore normal function. While devices are typically designed and used exclusively for one of these two purposes, there have been increasing efforts in developing implantable electrode arrays capable

About the Speaker

- President of the International Union for Medical and Applied Bioelectrography
- Bio-Well Co., Israel
- Research interests include Biophysics, wellbeing, biofield, human energy, water, anthropology, history
- Author or coauthor of 12 books, including: *The Emerging Science of Water* (2017), *The Energy of Health* (2017); and *The Energy of Consciousness* (2012)
- Published more than 200 journal papers; holds 17 patents
- PhD in Quantum Physics, PhD in Biophysics

of housing cultured cells, referred to as biohybrid implants. Once implanted, the cells within these implants integrate into the tissue, serving as mediators of the electrode–tissue interface. The biological component offers unique advantages to these implant designs, providing better tissue integration and potentially long-term stability. In the following sections, the author defines several main areas of practical applications of this approach.

1. Cell Transplantation in Regenerative Medicine

Transplanted stem cells can be used in two different ways to restore function. First, in cases where a particular cell type population has been severely depleted or damaged, stem cells can be programmed to differentiate into the said cell type of interest and implanted. In this way, the cell transplant integrates into the host to become new tissue. Alternatively, cells can also be implanted to provide a favorable environment for host cells, so that these host cells may proliferate to repopulate a lesion. One of the main challenges faced when developing cell transplantation therapies is the risk of immune rejection.

2. Biohybrid Implants

These two different research avenues—culture and electrical probing of cells on multi-electrode arrays (MEA) *in vitro* and cell transplantation for tissue regeneration *in vivo*—can yet be combined into one concept: implantable electrical–cellular hybrid devices. Though their characteristics can vary depending on the context for which they are developed, biohybrid implants in essence consist of implantable electrode arrays containing live cells grown in culture. Once implanted, these cells integrate into the host tissue (either allowing the host tissue to grow into them, or growing into the host tissue themselves), forming a bridge between electronics and host. The cells of the implant therefore serve as mediators of the electrode–tissue connection.

3. Central Nervous System

The biohybrid strategies include technologies that incorporate both cells and electronics. By bridging the gap at the site of injury with a biohybrid approach, cells could act as biological amplifiers, transducing signals from electrical to biological, allowing reestablishment of lost connections. There are two key strategies that have been described in the literature targeting the CNS using this biohybrid approach: those with cells on electrodes and those with cells growing through electrodes.

4. Peripheral Nervous System

A traumatic lesion or degenerative neuropathy to a peripheral nerve can lead to axon degeneration and destruction of the neuromuscular junction (NMJ). Unlike the CNS, the peripheral nervous system (PNS) has the capacity for regeneration, however, lesions with nerve gaps over three centimeters have a lack of functional recovery due to the distance of axonal regrowth required to reach the target. Consequently, loss of communication and electrical connections occurs from one part of the nervous system to another, affecting motor control and sensory processes. When trying to preserve or regenerate this NMJ, it is important to have a stable interface with the peripheral nerve and electronics. A biohybrid approach incorporating cell therapeutic strategies with electronics could help solve this problem.

5. Special Senses: Hearing

It is common clinical practice to restore hearing in deaf patients using cochlear implants to electrically stimulate spiral ganglion neurons (SGNs). Acoustic energy is converted into an electrical signal through an external device.

Future Perspectives

Biohybrid implants lie at the interface of biology and electronics and constitute a unique tool with great potential in both the clinic and basic research. Despite the multiple challenges in their design and fabrication due to the need to integrate expertise in microfabrication, cell culture, and implantation surgery, amongst others, biohybrid implants offer numerous advantages over traditional, implantable electrode array designs. This niche held by biohybrid implants can translate into unique therapeutic opportunities. ◊

References

Giandomenico, S.L., S.B. Mierau, G.M. Gibbons, L.M. Wenger, L. Masullo, T. Sit, M. Sutcliffe, J. Boulanger, M. Tripodi, E. Delivery, O. Paulsen, A. Lakatos and M.A. Lancaster. 2019. "Cerebral Organoids at the Air–Liquid Interface Generate Diverse Nerve Tracts with Functional Output." *Nature Neuroscience* 22: 669–679.

Obien, M.E.J., K. Deligkaris, T. Bullmann, D. Bakkum, and U. Frey. 2015. "Revealing Neuronal Function through Microelectrode Array Recordings." *Frontiers in Neuroscience* 8: 423.

Perin, R., T.K. Berger, and H. Markram. 2011. "A Synaptic Organizing Principle for Cortical Neuronal Groups." *Proceedings of the National Academy of Sciences of the United States of America* 108 (13): 5419–24.

Petit, G.H., T.T. Olsson, and P. Brundin. 2014. "The Future of Cell Therapies and Brain Repair: Parkinson's Disease Leads the Way." *Neuropathology and Applied Neurobiology* 40 (1): 60–70.

Rochford, A.E., A. Carnicer-Lombarte, V.F. Curto, G.G. Malliaras, and D.G. Barone. 2019. "When Bio Meets Technology: Biohybrid Neural Interfaces." *Advanced Materials.* https://doi.org/10.1002/adma.201903182. Accessed on October 8, 2021.

Session 2

Discussion

Harald Walach (Session Chair): Patrons, we now have half an hour of discussion, and the floor is open for questions. May I suggest that we focus first on the first talk?

Jonathan Schooler: I have a question.

Walach: Yes.

Schooler: Arnaud, that was a fascinating talk. I wonder if you could possibly talk about more conventional explanations for, say, the judgments about the photographs. For example, could people possibly be picking up on the health characteristics suggested by the physiologies in the photos? Or perhaps even some personality characteristics that might suggest the kind of death they might be susceptible to?

Arnaud Delorme: Yes, that interpretation is totally possible, and we discussed it at length. The idea is that when you show a picture of someone, actually, if you have twins, one is smoking and one is not smoking, you can tell with very high confidence which one is smoking, just based on their picture. We tried to control the image as well as possible, and these were images from school pictures—images taken 50 years before the people passed—and people had to guess which one is deceased and which one is alive. It is possible that even if the picture was taken a very long time ago, there are some subtle cues where you can tell that this person is going to die sooner than this other person. We tried to compensate for that. Now, you have more obese people in one category than another category, because this is a source of morbidity, but it is definitely a possibility.

In a second experiment, we asked, how did these people die? Because you could say that specific traits, such as personality traits, could be potentially reflected in fashion and facial features. This could lead to them being more impulsive and having a higher likelihood of getting killed by a firearm. So, the whole idea is that there might be very subtle facial features that mediums could pick up on and improve their performance, and we cannot rule that out—and this is an interpretation of the data. Because we have spent so much time controlling for them, we have said that these conventional explanations are not necessarily the most likely explanation, but they definitely provide an explanation.

Walach: Thank you. The next question is by Didier Guignard.

Didier Guignard: My question is for the first commentator in the first part of the session. It is about the three dimensions of time. I completely agree with his viewpoint that the "block universe" is something that does not correspond with reality. But he then showed a diagram with some image that is usually used for the Everett concept of time in connection with quantum physics. But the Hugh Everett view is leading to a multiverse in which there are several versions, and it is not like a choice. It still supports the idea of determinism, which is included in the block universe. It is just a different way of maintaining determinism against free will, or chance, or whatever. So, I would like to ask him, what is his interpretation of this? Does he see it as a real choice or as this multiverse view that Hugh Everett had? Thank you.

Walach: You want to go down that line, Jonathan?

Schooler: That is a very astute and provocative question. I think there are multiple ways in which you could implement the model I was suggesting. In one way, we could be, rather than the victim of being stuck in a single block universe, be the victim of purely chance outcomes, which drive us down one branch versus another branch. I think that would be a defensible perspective. But once you conceive of the possibility of there being alternative possible directions in which each of us could move, it raises the possibility, at least, that consciousness itself, which we really do not understand, could have some causal role. So, the idea that I have been contemplating is that, through interpretation and possible other mechanisms, we may have some impact on which branch we go down. Then this would be consistent with suggestions of random number generator studies, which have suggested that will and interpretation might influence which alternative outcomes come about. The bottom line is that, once we recognize that we do not understand how consciousness fits into the physical world, it at least raises the possibility that it may play a role that we do not currently appreciate.

Walach: Thank you. Kim Jin Choon will ask the next question, and then Serge Kernbach.

Jin Choon Kim: I have two questions. In my understanding, the concept of the theory of Korean Professor Won H. Kim is related to quantum theory. In other words, the solution of the Schrödinger wave equation has an imaginary part. So, this theory is related to the imaginary part of the wave function, and this is somehow related to Dr. Kernbach's biohybridizations, because he emphasized quantum biology. So, this theory of biological beings or objects somehow has a relationship with biohybridization. What is the relationship between those two concepts? My second question is, how can we apply those concepts to environmental problems? Because the topics of this second session are about solving environmental issues, how can we understand or apply the concept of the theory and biohybridization to resolving environmental issues?

Walach: This one is for Professor Won H. Kim and for Professor Kernbach. Professor Kim, are you with us? He is not online. So, Professor Kernbach, maybe you want to take that first?

Serge Kernbach: Yes. I will try to respond in more general terms, though not exactly related to the point by Professor Won H. Kim. With biohybrid interfaces, we discovered a lot of very different issues, some of which are very controversial. Of course, we try to find some explanations and provide some hypotheses. Currently, in the area of quantum physics, there is a trend in that whatever we do not understand we say, "It is a quantum effect." We should, of course, avoid this. Quantum biology provides an integrated look at different phenomena. The quantum explanation for the oldest phenomenon, the phenomenon of light, is probably a good way to approach this. It is kind of a general answer, if I understand the question correctly.

Walach: There was also a question as to whether this would help us solve the environmental problem. Could you answer that?

Kernbach: Yes. In the case of environmental problems, we have a need to solve some of them immediately—these are problems that we should probably solve with more direct means. If the need to solve them is more distant, or over some time, then the way of quantum biology is probably the next step in solving the problems. As soon as we understand all these quantum phenomena in a macroscopic system, if we can understand entanglement effects, then the next step is the generation of technology. If technology appears, then it can essentially contribute to the environmental issues. By the way, may I ask a question?

Walach: You can ask your question because Professor Kim is currently not online.

Kernbach: The question is for the first speaker (Dr. Delorme), about your experiments with sensitive people. We, too, are doing this type of experimentation, using the "Maharishi effect." Many, many people are involved in this kind of experiment—it was on Facebook with hundreds of people. Sometimes, you see a clear difference between the individual response of one person and the collective response of many persons—what would be called "collective consciousness." My question for you is, did you observe any kind of collective phenomenology in exploring consciousness issues?

Walach: Arnaud, that was for you.

Delorme: I did not get the last part of the question. Could you repeat it?

Walach: He asked whether there was a collective type of phenomenology that you could observe.

Kernbach: In this consciousness issue.

Delorme: Collective phenomenology?

Kernbach: Yes. When you invite, let us say, 12 persons, and you explore them individually, there is a kind of individual response that you are exploring. But if you invite 12 persons together and they start to interact with each other—when they enter a single room, they start some interactions. Did you explore this kind of collective phenomenology?

Delorme: In our experiments, we try to be double blind. For example, when we ask questions, the person who asks questions does not know that the other person is deceased. So, there cannot be some kind of conscious or unconscious information transferred. We call that type of information transfer a "cold reading." When a medium comes, if they have a client who is in front of them, then they might detect some stubble cues from the client and acquire information about them. So, we try to minimize that. That is why we try to minimize the influence of the experimenter. So, in this case, I do not think that that would happen. But in the case of a standard reading, that is definitely a confound. I do not know if you are suggesting that we might access collective consciousness. Again, that is definitely an interpretation. As I mentioned in my presentation, the main hypothesis is something that we do not understand what is going on. So, it is not necessarily trying to characterize it, but we do not understand it. It could be collective consciousness, or it could be connecting to deceased spirits. We do not make any conclusion on that point.

Walach: Okay, thank you. Now, Professor Kim, are you online? Can you hear us?

Won H. Kim: Yes. I am sorry, I had some audio problems. I want to say something about water memory and imaginary space. Because a new paradigm arises from an unexplainable but very reproducible fact, I think the best candidate is water memory. Scientists are trying to explain water memory with a coherent domain of [Emilio] Del Giudice or EZ water of [Gerald] Pollack. EZ water may be a coherent domain that could be an efficient medium for water memory. But I want to ask a basic question about what is memorized to water. I hypothesized that the imaginary part of a substance can be separated from the substance and makes a 3D wave in EZ water or in the coherent domain. I want to note that water memory is not confined to water. I found that the 3D wave of a substance could be imprinted onto another medium, such as crystals or ceramic balls, and furthermore, it could be digitized and expressed in flat, two-dimensional space, and it could even be modulated to electricity. This 3D wave hypothesis can explain water memory, and it shows that even digitization of water memory is possible. I tried digitization with a visual image and can print it out in flat 2D space, for example, in a card. A digitized 3D wave commands imaginary space to form a 3D field of the substance around a card.

Dr. Arnaud Delorme's presentation could be explained with imaginary space because it is explained by a hologram. A hologram is basically a dialogue between a part and the whole. We know by experience that holograms exist, and universes and life are holograms, but that cannot be explained by a materialistic viewpoint. So, how can we explain holograms? I have proposed that imaginary space could explain a hologram, because every physical space is connected—it has an imaginary part. Imaginary space does not have space and does not have time, as it is basically a hologram space. Every piece of matter is connected to imaginary space. So, every physical space, every piece of matter in the universe is connected to the others.

Walach: That was a great answer, and we need to move on. So, I want to ask those present whether there is still another question for the first talk and its commentators. Arnaud, do you want to ask something?

Delorme: This is a question for Jonathan. I like the idea of different levels of consciousness encapsulated into each other. Then, how would he explain that a given level is not conscious of the lower levels, and how are they in synchrony with each other?

Schooler: Thank you very much for that question. The idea is that you would have synchronization happening essentially like carrier waves, in which you have a lower level of waves that have a particular rhythm, and they are synchronized, or cross-coupled, with the higher levels. In the same way that you can have very high frequency waves or vibrations synced in with lower frequency ones in a sort of cross-coupling manner, so too, you could have the rhythms of the lower-level ones connected up to the higher-level ones. Through cross-frequency coupling, you could potentially have it going both top down and bottom up. We have a paper that we just submitted to the journal *Behavioral and Brain Sciences* that articulates this in more detail, and I would be happy to share it with anyone who wants to reach out to us.

Walach: Finally, the problem is solved! Thank you, Jonathan. I suggest we move on to the second presentation, that by Professor Serge Kernbach on biohybridization, and that people raise their questions or make comments. If there are none, let me begin. I find that area very fascinating because, as he said, it is a huge industry, and industry always drives development. My question would be, is there any ethical guidance, and if so, what is it?

Kernbach: Yes, there are a lot of ethical limitations, for instance, there are issues with performance in European projects, and even experiments with animals, with fishes. There should be a lot of improvement, through ethical commissions and universities where talks are usually given. From this point of view, I think there will be even more in the development of technology with plants. We work mostly with plants. It is much easier to perform these experiments and develop the technology with plants. One of the essential steps is the connection between plants and technology, or ecosystems and technology. That is what we are doing.

Walach: Can we implant an electrode into a soldier, say, and remote control him to do stuff that he would probably not do if he were not controlled?

Kernbach: Do you mean with plants?

Walach: No, not with plants but with brain interfaces, for instance.

Kernbach: With brain interfaces? I think there are a lot of second-generation ethical issues, such as the issue of biological or digital immortality. If, let us say, some super-rich person is "immortal," we can think about the consequences for societies, or some people who are considered "mighty" from a social viewpoint also may remain frozen "forever". The implanting of electrodes in the brain is the next point. One question is whether the technology is hackable, because we know there are a lot of hackers and hacking of biohybrid devices. We now know about the hacking of mobile phones, and all these personal movies get public somehow—it is extremely unpleasant. Or some data get stolen. But if the implants, for instance, a payment chip implanted under the skin), would be hacked—it is, of course, absolutely unpredictable what can happen. I mean, there is some deficiency that we probably have to pay more attention to on this development.

Walach: Well, yes. I think there is a question by Arnaud?

Delorme: My question is with respect to robots and artificial intelligence and micro electrodes. I am sure we all know of Elon Musk and his company, Neuralink. His long-term goal is to transfer one's consciousness—either he wants to be immortal or he wants to transfer his consciousness to a machine. So, I just wanted your opinion on this. In the long term, even looking forward 100 years from now, do you think it is possible, or are there ontological barriers that make it impossible?

Kernbach: Yes, for instance, my personal dream is to have, instead of huge mobile phones, something like a really small chip integrated into the body. Then I would have access to all this knowledge of human society, and all these communication capabilities. It is a first step, and I think it is already achievable; probably in 10–15 years, it will be fully achievable. The next point is if we can completely transfer our consciousness to these devices. We need to understand what consciousness is. If consciousness has only biological embodiment, then, of course, we can probably digitize it and then store it in a cyber simulation, like Microsoft has more or less proposed, it would probably be feasible. But the question is, what if consciousness is not restricted to biological agents only? If we have some kind of parallel machinery, such as quantum or something else, then the big question is whether we will ever be able to do all this digital immortality stuff. From this point of view, I think we first have to gain more knowledge about our consciousness.

Walach: Well, maybe the final question is, what really is consciousness? Is it just what the brain produces? Or is it more than what the brain produces, and the brain is just some kind of relay mechanism? Apparently, we do not know that yet, do we?

Kernbach: Yes, correct. We have several hypotheses. In hypotheses from the '70s and '80s, there are only biological ideas about consciousness as a kind of computational biochemical machinery, more or less. Now, I think the common understanding in neuroscience is that we probably have other kinds of effects—these quantum effects in neurons. Many papers have been appearing from around 2019 and 2020, that say, let us revise this hypothesis and say that biological computation includes some kind of quantum phenomena, for instance, in the proton 2 million cafes, or in neurons, there is a lot of quantum computation going on. From this point of view, my assumption is that consciousness is a little bit more complicated. So, we definitely have biological embodiment, but now it seems more and more probable that we have nonbiological embodiment as well—whatever the kind of nonbiological embodiment, whether it is through weak signals of some sort, or a quantum region, or a combination that people call a "biofield." We do not know, we can only assume, but it seems that we have two embodiments. This is our current understanding.

Walach: That sounds to me like a very good final word for that session, because our time is up. I think we need to close, even if we could not answer all the questions that could be asked. So, let me thank all of you—all the speakers and commentators, and say goodbye. Thank you. ◊

ICSG II

SESSION 3

Ramifications of the Postmaterialist Perspective

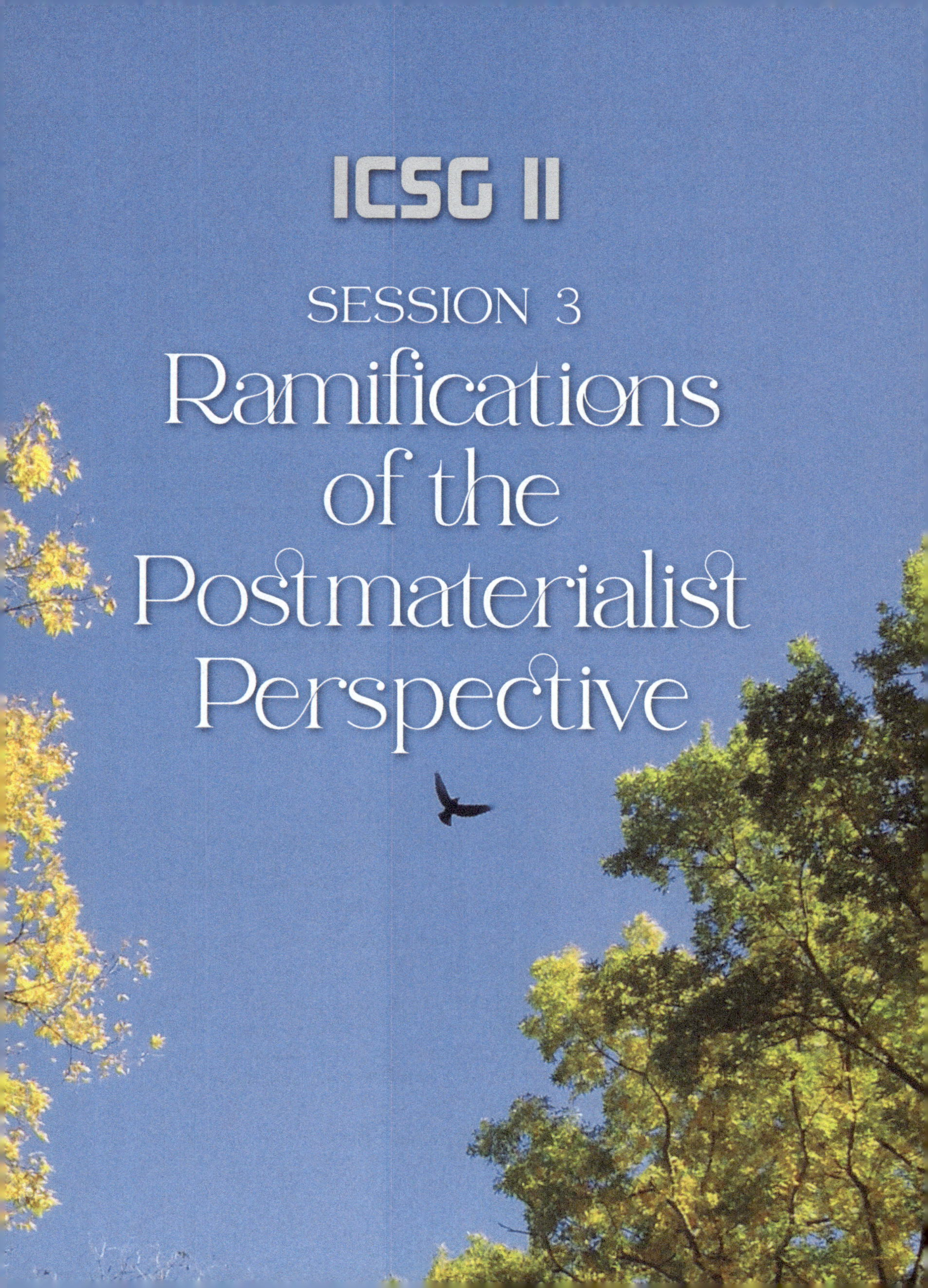

About Session 3

Ramifications of the Postmaterialist Perspective

Scientists who take a materialist perspective believe that human consciousness, or the "mind," is a product of the brain, which is part of the material realm. From this viewpoint, when a person dies, the brain stops functioning and the person's consciousness disappears. On the other hand, from a postmaterialist perspective, the universe also has one or more intangible, nonphysical realms, and human consciousness exists in a nonphysical realm and persists beyond physical death. The postmaterialist perspective is supported by investigations in frontier science, which have opened the door to invisible dimensions of reality. Thus, we may say that human thoughts (or consciousness) exist in a nonphysical realm, while the human brain, the body, and the natural environment exist in a separate, physical realm. However, there are mutual interactions between our consciousness and our body and the environment. In this session, one talk proposes an explanation of such interactions. The other presentation shows how such investigations could lead us to a better understanding of the invisible ultimate cause of the universe. This understanding could inspire personal and environmental restoration. ◇

About the Chair, Professor Menas Kafatos

- Fletcher Jones Endowed Chair Professor of Computational Physics. Chapman University, Orange, USA.
- Director of the Center of Excellence in Earth Systems Modeling and Observations (CEESMO)
- Author or coauthor of books in many languages, including: *The Conscious Universe; The Non-local Universe* (with Robert Nadeau); *Principles of Integrative Science* (with Mihai Draganescu); *You Are the Universe* (with Deepak Chopra); and *Science, Reality & Everyday Life.*
- PhD in Physics, Massachusetts Institute of Technology

Session 3 | Presentation One

How Does Human Consciousness Interact with the Brain, and God's Consciousness with Human Consciousness?

Dr. Glen Rein
Founder and CEO, Quantum-Biology Research Lab, Boston, USA

First, I would like to thank the organizers of the conference for the invitation to speak at this awesome conference about science and God. Today I would like to talk about how human consciousness interacts with the brain, and how God's consciousness interacts with humans.

I would like to start with the concept of the mind of God, because to me this sounds like an oxymoron. **Figure 1** shows a quote from Stephen Hawking,

About the Speaker

- Founder, Quantum-Biology Research Lab, Boston, USA. Funded by the Fetzer Institute, the lab has pioneered the study of biological effects of nonclassical electromagnetic (EM) fields
- Discovered and published a new effect of neuropeptides on immune cell function
- Performed traditional biomedical research at Harvard and Stanford Medical Schools, involving studies in neuroscience, immunology, psychoneuroimmunology, and bioelectromagnetics
- Published in *Nature* the first demonstration that EM fields modulate neurotransmitter function *in vitro*
- Co-developed a new electro-diagnostic method to measure endogenous EM fields of the body
- PhD in Neurochemistry; PhD in biochemistry from the University of London

THE MIND OF GOD

Did God cause the Big Bang?

Does God play dice?

The unknowable

The infinite

"If we discover a complete theory... then we are able to discuss why it is that we and the Universe exist. If we answer that... it would be the ultimate triumph of human reason, for then we would truly know the mind of God."

— *Stephen Hawking*

"I want to know the mind of God, the rest are details."

— *Albert Einstein*

Figure 1

who said that "if we knew the ultimate reason for the existence of the universe, then we would know the mind of God." There is also a quote from Albert Einstein, who wanted to know the mind of God, the rest being details. I, however, believe the phrase "the mind of God" is a non-sequitur. I believe that God does not have a mind or a brain or a body. For me, God is energetic and not physical, so God cannot have the attributes that humans have.

If we look back at the history of this concept of the relationship between the brain and the mind, or the brain and consciousness (**Figure 2**), we can see it started in the time of the Greek philosophers who speculated that the mind and therefore assumedly consciousness is nonphysical, but it interacts with the brain. That notion was a bit of a paradox for them, because it is not obvious how something nonphysical can interact with something physical. To me, they were actually right: the mind is nonphysical. Therefore, according to Einstein's equation, if it is not physical, and it is not matter, it must be energy.

So, I have developed this simplistic model (**Figure 3**), where we show God on the far left, human consciousness or individual consciousness of

HISTORY OF MIND-BRAIN DUALISM

- Greek philosophers – soul is non-material – separate but part of brain
- Hippocrates – brain is interpreter of consciousness (CON) – interactionism
- Descartes – non-physical mind interacts with brain soul interacts with pineal
- Darwin – thought is secretion of the brain
- Morgan – 1920s – mind and CON are emergent properties of the brain
- Freud – mind – preconscious, conscious and unconscious
- Spinoza – mind=body, both attributes of God

If mind is non-physical, it is energy $E=mc^2$

Figure 2

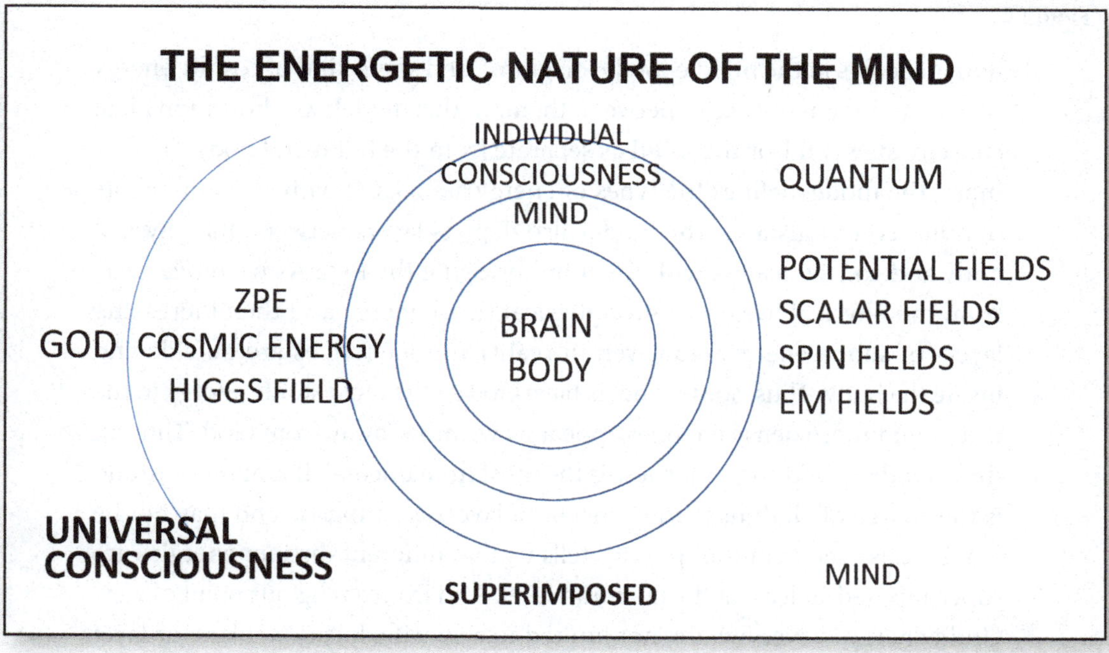

Figure 3

NEURAL BASIS OF CONSCIOUSNESS COGNITION

- Computational neurobiology input vs output
- Information processing within the brain
- Qualia – conscious experience = consciousness?
  - – separate from brain?
  - – exerts control over electrophysiology of brain?
  - – arise from brain activity?
- Encoding information from within the brain
- Encoding information from outside the brain
  - Vision
  - Electromagnetic Fields
  - God

Figure 4

human beings in the middle (of the diagram), and some kind of cosmic energy or zero-point energy (ZPE) between them. In this model, we show individual consciousness and/or the mind as separate from the brain and body. Furthermore, the model defines the types of energy associated with or components of mind/consciousness. The model also depicts layers between the physical body and consciousness with the mind bridging the two. As we move away from the physical, we encounter physical (EM) energy and then increasing layers of subtle energy. However, in reality these energies are outside and inside the body. Thus, some models have God as the most subtle energy in the center and more dense energies appear as we move away from God. Thus, in these models, God consciousness is the most fundamental. It is at the core and is the source of all things. The concept of layers is simplistic and may not be real because the quantum physics tells us that different distinct energies are superimposed, at least at the quantum level. And from a religious point of view God's energy is everywhere, not limited to a specific location, plane or layer.

Going back to the simplistic model in **Figure 3**, on the far right I have laid out several types of energy fields starting at the quantum level (which is the most fundamental) all the way down to electromagnetic (EM) fields, which are the most dense. This is one way of characterizing the energetic nature of the mind. Electromagnetic fields are primarily in the physical body because we know that these fields are generated from the nerves in the brain and they can affect the brain and the body. Thus, EM fields are highly associated with the body, whereas the three types of subtle energies are associated with the mind.

In modern times, the focus has been on the neuronal correlates of the mind/consciousness (**Figure 4**) and not on the energetic nature of the mind. Most contemporary scientists are still saying that the mind and brain are separate. In fact, contemporary models address the brain (and not the mind) addressing topics such as the information capacity, information processing, and the computational capacity of the brain. In the scientific literature it is not clear how these concepts extend to the mind or the conscious experiences we all have, which are called qualia in today's terminology. There is still a need to distinguish between the physical brain and the energetic entity, the mind.

1. The Holistic Level of Brain-Mind Interaction

The details of exactly how information is encoded within the brain is still not clear, let alone how information encoded outside the brain gets inside the brain? That is where God comes in. Many people think that God is out there somewhere, outside of us, in the sky somewhere—at least many religious people think so. The more scientifically inclined people tend to put God at the center or core, as presented above. Since the subtle and EM energies described above are part of the brain, is God also part of the physical brain? For example, in the case of vision, light and the information come from outside the brain into the retina and EM fields in our environment are known to impinge upon the physical brain. Do these energies also effect the mind? Can the mind (via intention) effect subtle and EM energies? If so, can God's energy effect our physical bodies and our mind? Can our mind effect God or "resonate" with God at all? That could be the scientific basis for prayer.

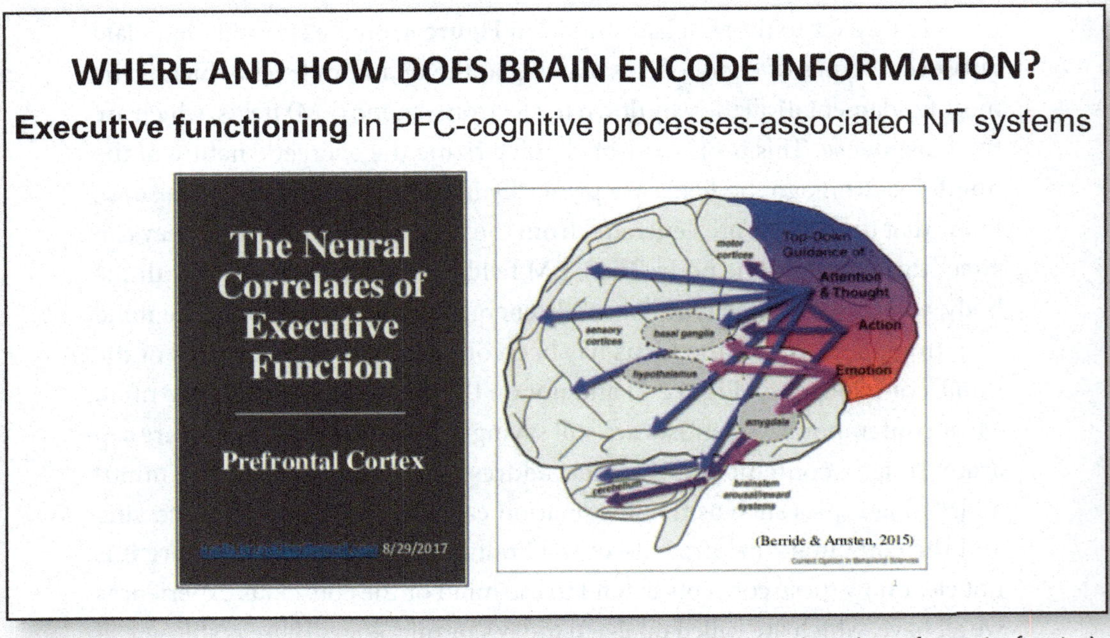

Figure 5 (Source: https://www.slideshare.net/OzellaBrundidge/the-neural-correlates-of-executive-function)
(Original study: https://www.sciencedirect.com/science/article/pii/S2352154615000121?via%3Dihub)

When talking about the neural correlates of consciousness, let's start with a global perspective and work our way down to the molecular level. According to one model (**Figure 5**), it is all about the executive function of the brain and the prefrontal cortex. This model proposes consciousness (and perhaps the mind) is localized in the prefrontal cortex and from that location can influence all the other parts of the brain. This model seems rather simplistic and it is not clear why the prefrontal cortex is given such a fundamental role.

Other models have a more global perspective focusing on neuronal networks. Neuronal networks are patterned arrays of complex mixtures of different individual neurons. Neuronal networks can transfer information back and forth within a given network, thereby allowing for nonlocal interactions between individual neurons (**Figure 6**). Assuming different networks can communicate with each other, consciousness can be considered the global interplay between all networks in the brain. Needless to say, this would be difficult to validate experimentally.

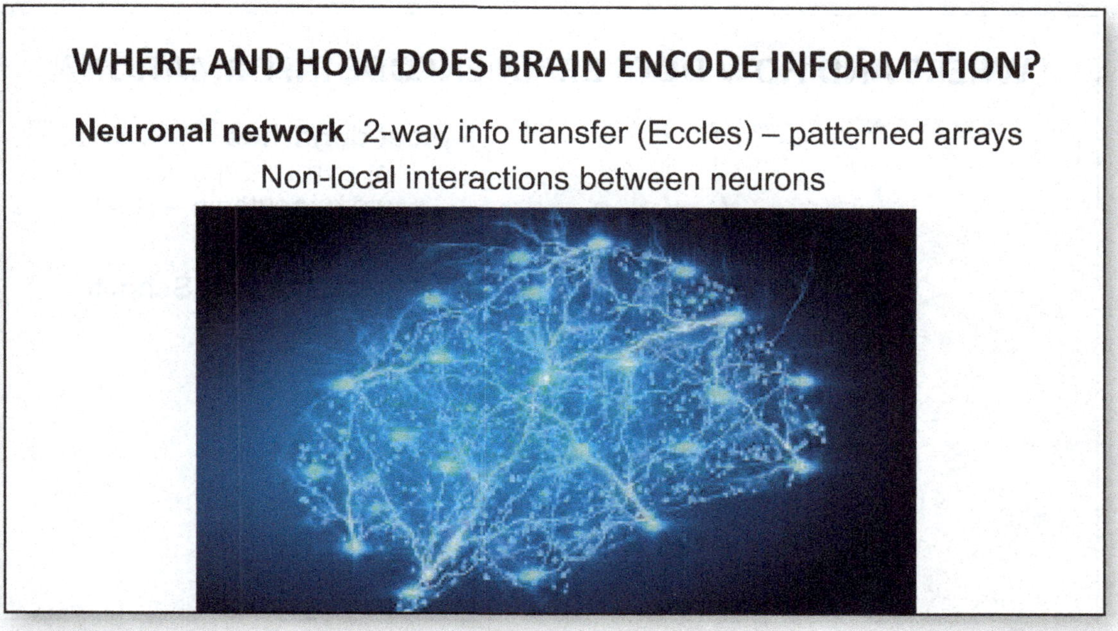

(In webpage: https://nautil.us/surprising-network-activity-in-the-immature-brain-10925/)
(Image source: https://assets.nautil.us/nautilus-vertical/maxplanck-a-354.png?auto=compress)

Figure 6

2. The Cellular Level of Brain-Mind Interaction

Some models focus on the neurons themselves. One particular model, first introduced in the 1960s by Beck and Pribram (**Figure 7**), proposes a fundamental role for pyramidal cells in the cerebral cortex. It is proposed that the communication between their apical dendrites and their basal dendrites is linked to consciousness (Phillips 2016). After excitation, dendritic nerves can modulate how they respond. This decision-making process, referred to as neurocomputation, is therefore associated with consciousness.

The apical dendrites have been written about and used as a model for consciousness by other authors (**Figure 8**). A study by LaBerge (2007) proposed that columns which are formed from multiple dendrites are involved in consciousness because dendrites exhibit recurrent circuits when excited. Such neuronal circuits are based on cyclical firing of individual pyramidal dendrites. Such a system can keep recirculating even in the absence of input signals, and

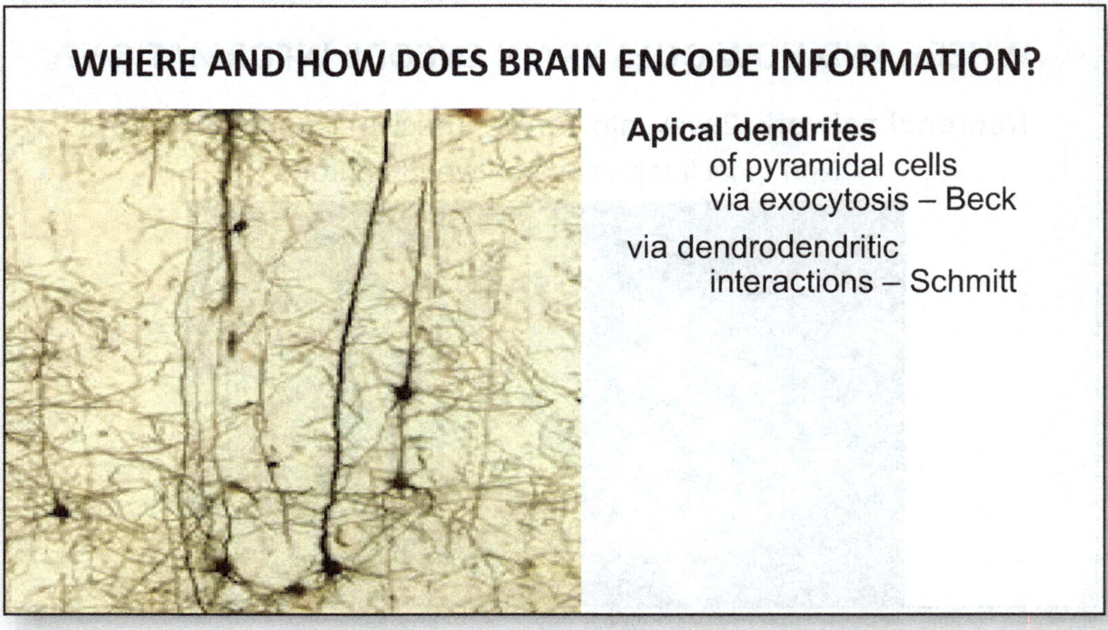

Figure 7 (Source: https://www.slideshare.net/OzellaBrundidge/the-neural-correlates-of-executive-function)
(Original study: https://www.sciencedirect.com/science/article/pii/S2352154615000121?via%3Dihub)

perhaps being modulated by input signals from consciousness or even God himself.

My own EM loop hypothesis (Rein 2020) introduces a new neural correlate of consciousness in terms of two interacting systems: the cyclic nerve firing of subcortical neurons within a neural net and pyramidal apical dendrites in the cerebral cortex. The two systems communicate with each other via EM (and subtle) forces and both systems can send and receive information. The interaction between the two systems exhibit collective ensemble behavior, which can be multi-dimensional with respect to space and time. The dynamic interplay and exchange of (quantum) information between the two systems becomes the interface which can be influenced by EM energy for sure, but also likely by subtle energy, and even God (Rein 2020).

The term "neuronal correlates of consciousness" is used to describe physical events in the brain that are associated with consciousness (or support it) (**Figure 9**). However, the exact relationship between the physical

APICAL DENDRITES THEORIES OF CONSCIOUSNESS

- **LaBerge** (2007)
  Minicolumns of dendrites in cerebral cortex contain recurrent circuits which underlie CON
  Attention activates AD in primary sensory areas (visual/auditory) which sends signal to higher sensory area generates elevated states of CON
- **Rein** (2020)
  "an EM loop mechanism whereby apical dendrites in the cerebral cortex can be activated from endogenous EM fields generated from neuronal networks to produce the reactions we describe as "consciousness"
  Internat J. Engineering & Science

Figure 8

NEURAL BASIS OF CONSCIOUSNESS

.... supporting consciousness
.... associated with consciousness

(Lower) Mind-Brain Dualistic Interactionism (2-way)
Functionalism: Mind is software algorithm w in/out functions
Higher Mind: higher cognitive processes
connected to spirit

Figure 9

brain and consciousness is still not clear. It leads us to the conclusion that there is a dualistic interaction between the mind and the brain because they both exist, and that information flows back and forth between the two. Since most scientist believe consciousness and the mind are the same, we can distinguish between higher mind and lower mind. It is the "lower mind," which communicates with the physical brain. It is in the higher mind where the higher cognitive processes occur, and the higher mind is the part of the mind that is connected to consciousness, to spirit, and to God.

3. The Sub-Cellular Level of Brain-Mind Interaction

Now, we can keep going down to even more fundamental levels and look at the sub-neuronal and sub-cellular components of the brain that are involved with consciousness, at least from a theoretical point of view (**Figure 10**). The hypothesis proposes that it is not the neurons themselves that are mediating consciousness, but it is some component within the neurons. This starts way back at the time of John Eccles, who postulated an undefined module within the neuron that was linked to consciousness. Karl Pribram, a professor at Stanford University, postulated a similar microstructure. Stuart Hameroff and Roger Penrose had been postulating that microtubules, which are components of neurons, are important mediators of consciousness (Hameroff, 1998). A fourth example is the work of Evan Walker (Walker, 1970) who felt that the synapse itself is linked to consciousness. This is a logical hypothesis considering the synapse allows nerves to communicate with each other, as depicted in the diagram. He proposed that some kind of quantum mechanical tunneling allows information to be transferred between the presynaptic neuron and the postsynaptic neuron. The idea of molecules and biological systems functioning at the quantum level is linked to consciousness has been pursued by several sciences since the time of Walker. The numerous quantum theories of consciousness is outside the scope of this article and has recently become a scientific discipline in its own right.

Figure 11 shows a diagram of microtubules, which are composed of the tubulin protein. Tubulin exists in two forms: the alpha and beta conformations.

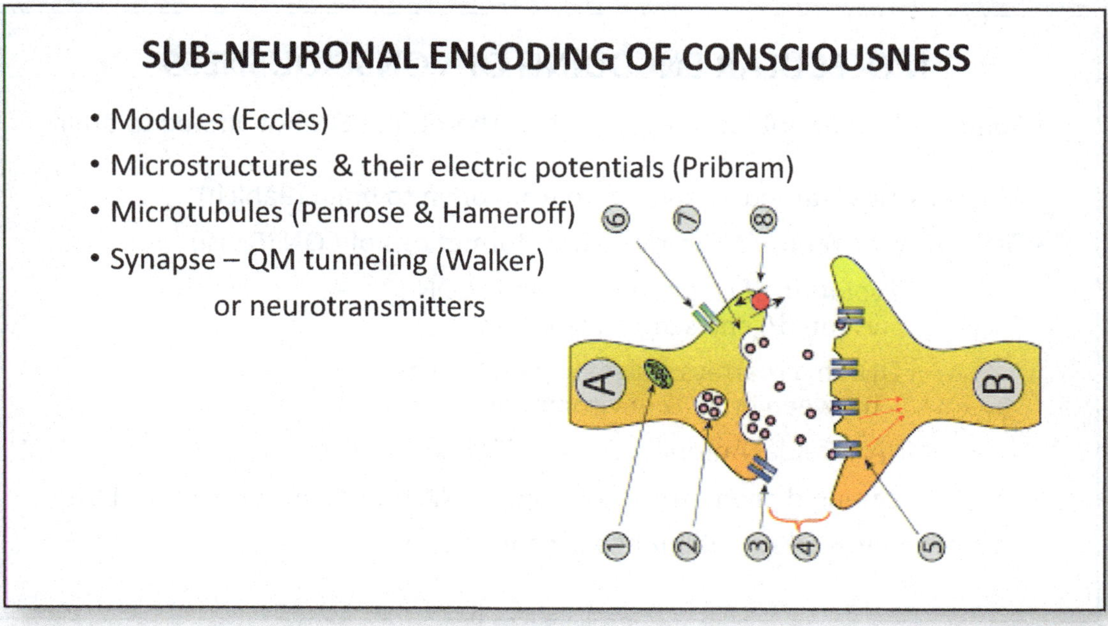

Figure 10

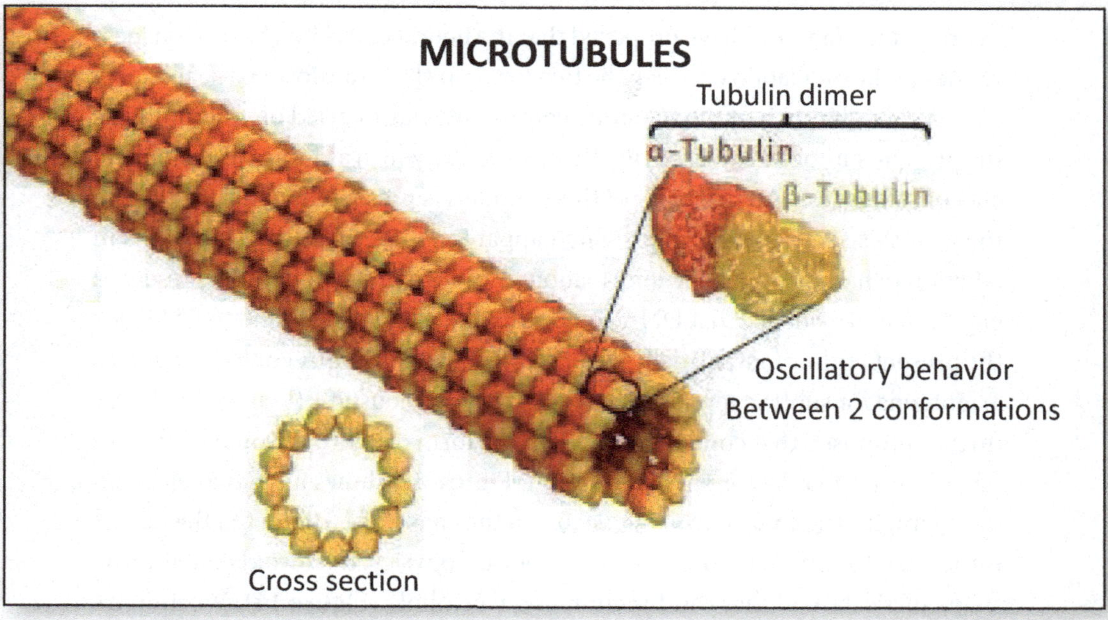

Figure 11

MOLECULAR ENCODING OF CONSCIOUSNESS

- Some molecules not localized in brain – Therefore CON not limited to brain
- RNA – carries learned & stored info from brain to brain (Babich)
- DNA – Psychogenomic Theory of Transformations of CON (Rossi)
  - Evolution of cognition to cosmic CON vs mind/gene search algorithms
  - DNA microarrays – signaling molecules between brain & environment
  - – DNA Toroidal Antennae Theory (Rein)
    - catenoid geometry win Toroidal DNA transforms EMF to NC EMF
- Porphyrin networks – exhibit phase conjugation

Figure 12

Penrose and Hameroff have proposed that the neural correlate of consciousness resides in the oscillation between the two forms as they transform back and forth.

Although much of the scientific community has focused on microtubules, the present author has focused on other molecules which as just as likely to mediate consciousness. An overview of these models is presented in **Figure 12**. The focus in this section is on DNA which apparently has additional functions in addition to harboring genetic information. Although previous authors presented experimental evidence that DNA could resonate with and respond to EM fields, the present author first formally proposed the idea that DNA could function as an antenna for subtle energies associated with consciousness (Rein 1996). It was further proposed that condensed DNA in the form of a toroid would be the preferred form of DNA for such a function. **Figure 13** shows an actual electron micrograph of a physical DNA molecule in the shape of a toroid. On the far left is the ideal toroid. According to contemporary physics, the toroid contains the shape of a catenoid that can function as a wormhole (**Figure 14**). Wormholes

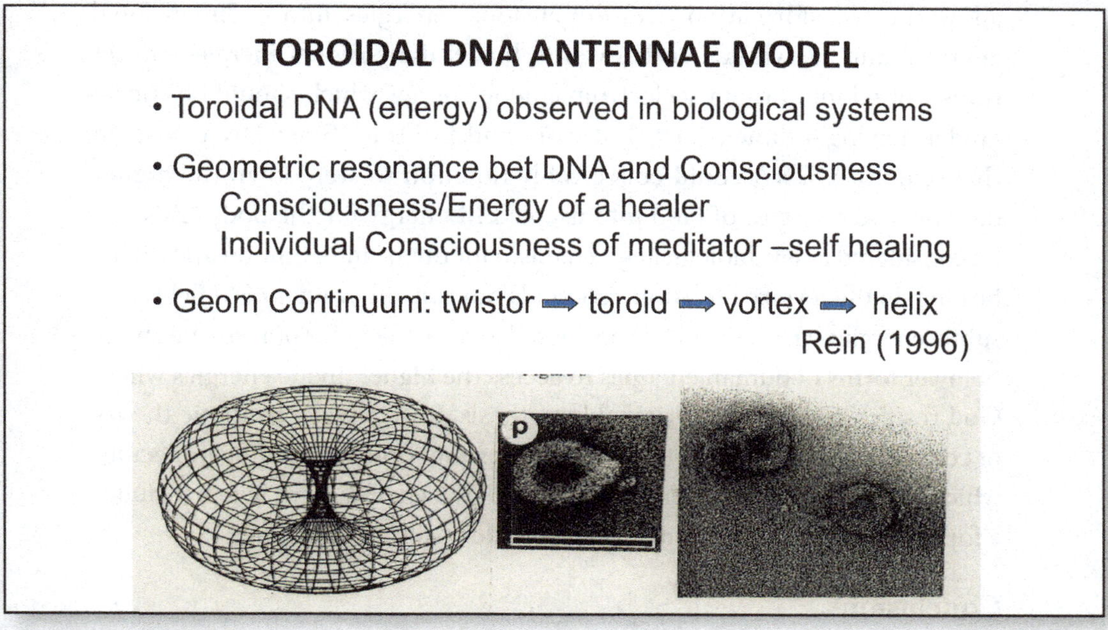

(Source: https://images.google.com or www.rsc.org/ibiology) **Figure 13**

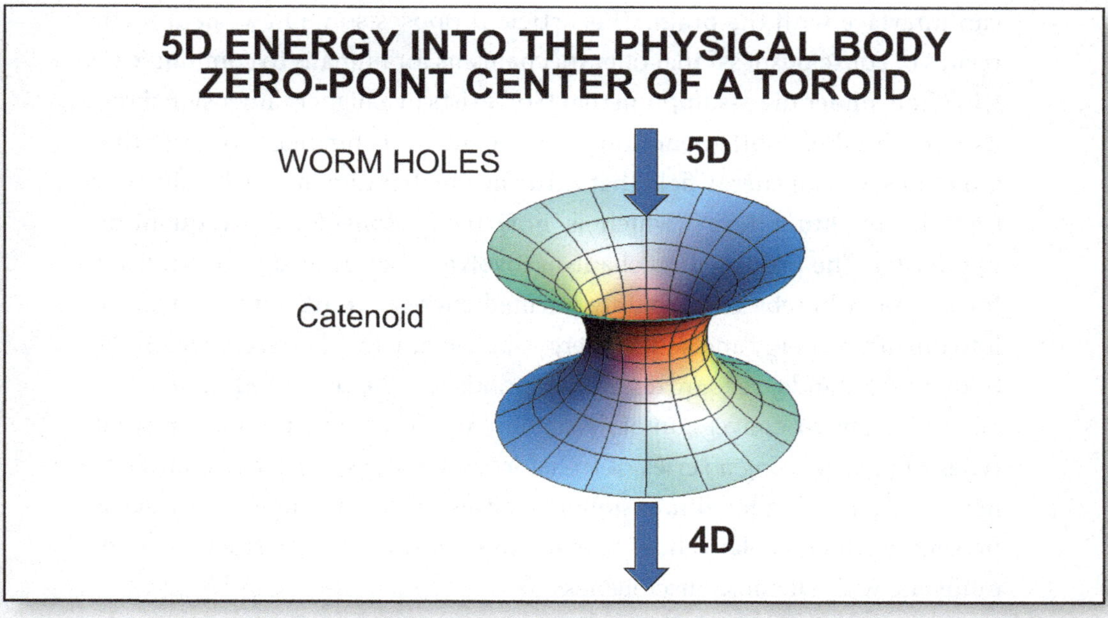

(Source: http://xahlee.info/surface/catenoid/catenoid.html Or https://i.stack.imgur.com/WVmog.png) **Figure 14**

allow the transformation of 5-dimensional energies into 4-dimensional energies, and vice versa. In other words, higher dimensional energies can get transformed into ordinary electromagnetic energy, which should be considered as having 4 dimensions, 3 of space and 1 of time. Since DNA exists in this structure, DNA could conceivably function to step down the higher dimensional energies of God into usable EM energy for the body. This can be considered a new biological mechanism for divine intervention and divine healing. In the reverse direction, catenoids can convert ordinary EM fields into subtle energies which could be considered a mechanism for prayer–the ability of lower forms of human energies to access the higher divine energies where God resides. Since Rein's original hypothesis, Rossi's pychogenomic theory of consciousness proposed that DNA arrays function as signaling molecules which allow the brain the communicate with the environment which includes a form of cosmic consciousness which could easily include God.

Conclusions

This article reviews previous hypotheses that speculate how consciousness can interface with the brain. This article proposes some new "neural correlates of consciousness" that offer mechanisms for humans to communicate with God, under the assumption that God exists in a higher dimensional reality composed of subtle quantum energy fields. It is further proposed that God himself is an energy field that exists in a higher dimensional reality beyond the quantum level (which is now being considered by quantum physicists). The proposed mechanism involves a series of step-down transformations whereby God's highly purified energy can be down converted into quantum energy and subtle energy and eventually EM energy which the body understands, recognizes and responds to. The proposed model also allows for the up-conversion of dense human EM energy into higher order types of energy using a series of transformation steps, until our consciousness can enter higher dimensional realities. Thus, a scientific model is presented which explains how human consciousness can interact and communicate with Divine Consciousness. ◊

References

Beck, F., and J.C. Eccles. 1994. "Quantum aspects of brain activity and the role of consciousness." In *How the Self Controls Its Brain* (pp. 145–165). Berlin and Heidelberg, Germany: Springer

Berridge, C.W., and Arnsten, A.F. 2015. "Catecholamine mechanisms in the prefrontal cortex: proven strategies for enhancing higher cognitive function." *Current Opinion in Behavioral Sciences* 4: 33–40.

Eccles, J.C. 1965. "Conscious experience and memory." In *Brain and Conscious Experience* (pp. 314–344). Berlin and Heidelberg, Germany: Springer

Hameroff, S. 1998. "Quantum computation in brain microtubules? The Penrose-Hameroff 'Orch OR 'model of consciousness." *Philosophical Transactions of the Royal Society of London. Series A: Mathematical, Physical and Engineering Sciences* 356: 1869–96.

LaBerge, D., and R. Kasevich. 2007. "The Apical Dendrite Theory of Consciousness." *Neural Networks* 20: 1004–1020.

Phillips, W.A., M.E. Larkum, C.W. Harley, et al. 2016. "The effects of arousal on apical amplification and conscious state." *Neuroscience of Consciousness,* Volume 1.

Rein, G. 1996. "Effect of conscious intention on human DNA." In *Proceedings International Forum on New Science,* Denver, CO.

———. 2017. "Mind as a Virtual Phase-Conjugated Hologram." *Cosmos and History: The Journal of Natural and Social Philosophy* 13: 217–226.

———, R. Callan, and J. Buekers. 2020. "The Electromagnetic Loop Theory – A New Paradigm in Consciousness Research." *The International Journal of Engineering and Science* 9: 15–19.

Rossi, E., and K. Rossi. 2014. "An evolutionary RNA/DNA psychogenomic theory of the transformations of consciousness: The quest for therapeutic mind/gene search algorithms." *The International Journal for Transformations of Consciousness* 1: 1–20.

Walker, E.H. 1970. "The nature of consciousness." *Mathematical Biosciences* 7: 131–78.

Session 3 | First Commentary
On the Presentation by Dr. Glen Rein

Science, Consciousness, and God: Closing the Epistemic Gap

Professor J. Scott Jordan
Department of Psychology, Illinois State University, Normal, USA

Most contemporary takes on the science of consciousness make strong distinctions between objective, physical reality and subjective, mental reality. As a result, the relationship between science and consciousness is one of determining how it is that an objective, physical phenomenon can give rise to, interact with, or cause a subjective, mental phenomenon such as consciousness.

While such an approach to the topic has led to amazing scientific findings regarding brain dynamics, including Dr. Rein's theory of quantum holography, objective-subjective divisions never seem to lead to an account that renders subjective, mental phenomena necessarily real. In other words, researchers

About the Speaker

- Professor and Chair, Department of Psychology; Director, Institute for Prospective Cognition, Illinois State University, USA
- Co-President, Society for Complex Systems in Cognitive Science. Teaching areas include: Learning and Cognition, Perception, Systems Theory
- Research interests include: Consciousness, Volition, Action-Planning, Intentionality
- Books edited include: *Intercorporeality: Emerging Socialities in Interaction* (2017)
- PhD in Cognitive Psychology and the Neurophysical Basis of Perception

propose theories in which all of the causal work goes on in physical systems while subjective consciousness rides along, as it were, never entering into a causal relationship that would render it necessarily real to scientific explanation.

Faced with such potential epiphenomenalism, one has at least two choices: continue to search for a recursive causal link between the objective and the subjective, or ask whether or not the objective-subjective distinction is actually necessary. While the former option dominates the contemporary scientific landscape, there is a strong history in Western philosophy of fleshing out the latter.

Specifically, philosophers such as Benedict de Spinoza, G.W.F. Hegel, Michael Oakeshott, and Maurice Merleau-Ponty avoided objective-subjective divisions that defined reality *dialectically*. That is, objective phenomena were being defined in terms of *not being* subjective phenomena, and *vice versa*. As a result, the very definition of objective and subjective negated the possibility of the two being of the same kind, while simultaneously necessitating the development of a causal bridge to cross the *assumed epistemic gap* between objective and subjective realms.

Hegel defined reality as an internally related unity, meaning that all things are about and constituted of all things (**Figure 1**). All objects are constituted of their subjects and vice versa. This is known as the doctrine of internal relations. And it is the idea that if you have phenomenon A and phenomenon B, and they share a relation, if that relation is constitutive of what A and B are, then A and B are internally related. Now, this idea of reality being this internally related unity challenges contemporary distinctions between the notion of subjective versus objective properties, or in more contemporary ways of thinking, the distinction between intrinsic versus relational or relative properties.

But does anyone really believe that reality reduces to context-independent, objective things that exist as if they are independent of all reality? Most people I asked, say "No" (**Figure 2**). Jammer (2002), for example, argues that mass, which is the archetypal intrinsic property, actually comes to a particle as it moves through the Higgs field. What is the Higgs field? Well, it is a field that permeates all of reality, which means that mass comes to a particle by

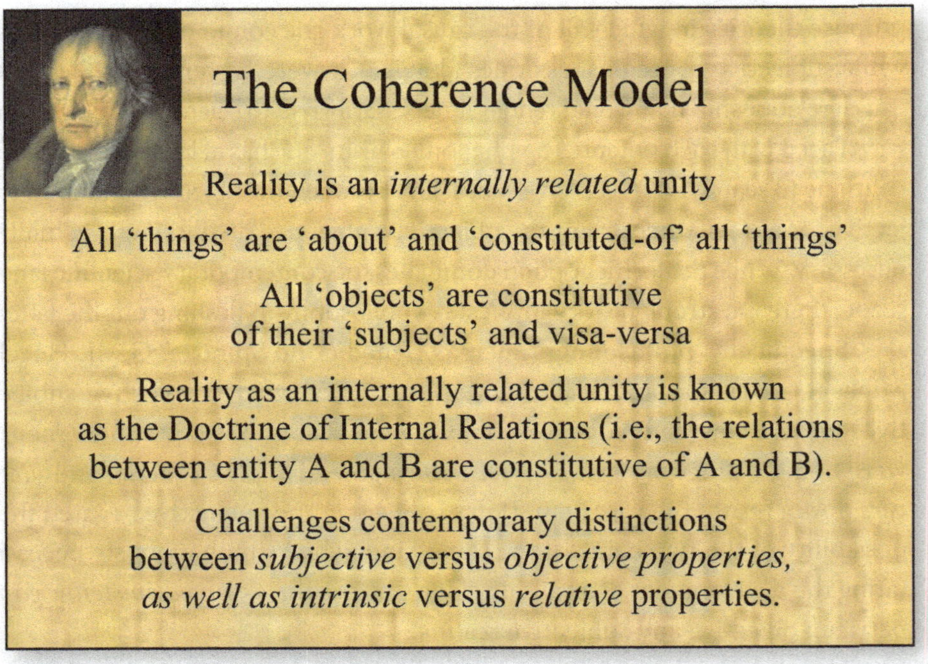

Figure 1 Image source: https://en.wikipedia.org/wiki/Georg_Wilhelm_Friedrich_Hegel

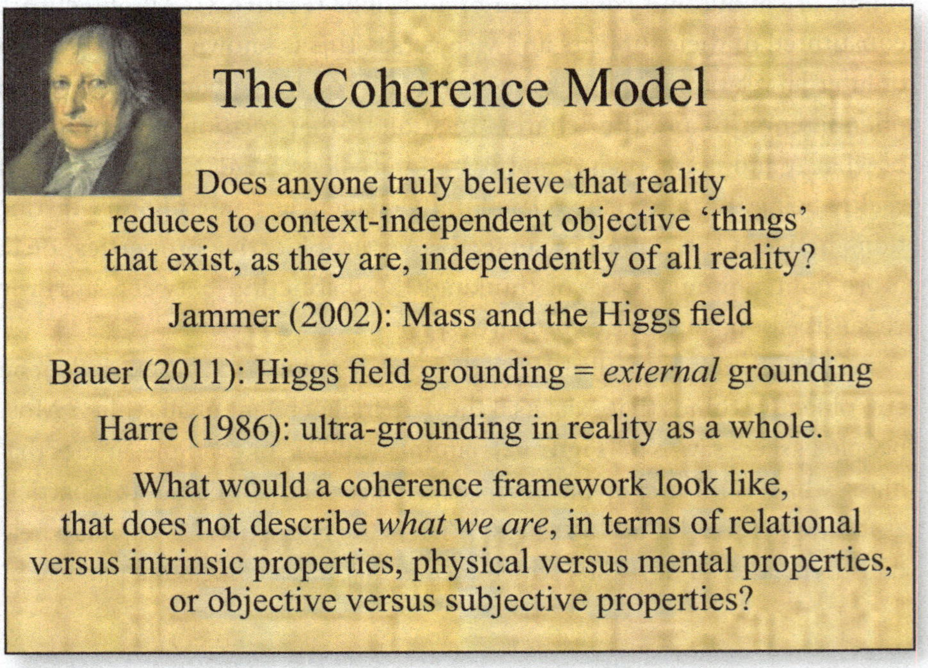

Figure 2 Image source: https://en.wikipedia.org/wiki/Georg_Wilhelm_Friedrich_Hegel

its interaction with all of reality, therefore, rendering mass a relational, not an intrinsic, property. Bauer (2011), commenting on this notion of the grounding of mass as a property coming through interactions with the Higgs field, calls this "external grounding." Then Harre (1986) puts forward the notion of "ultra-grounding," which is that all properties are grounded in terms of their relationality with reality as a whole. So, he takes this idea of an internally related unity and asks what it would look like to try to describe reality in terms of an internally related unity or thoroughly relational reality.

Dr. Rein (2016) takes an approach to consciousness that seems to suggest such a bridge when he asserts that light from the universal consciousness enters the top of a developing embryo's spinal cord and triggers a quantum hologram that ultimately constitutes an individual's consciousness (**Figure 3**). At the same time, like all theories based on objective-subjective divides, the hologram itself does not seem to share a causal relationship with the physical causality that generated it. As a result, the physical systems do all the work while the causally unnecessary individual consciousness rides along.

In my own work on Wild Systems Theory (WST) (Jordan 2013, 2020; Jor-

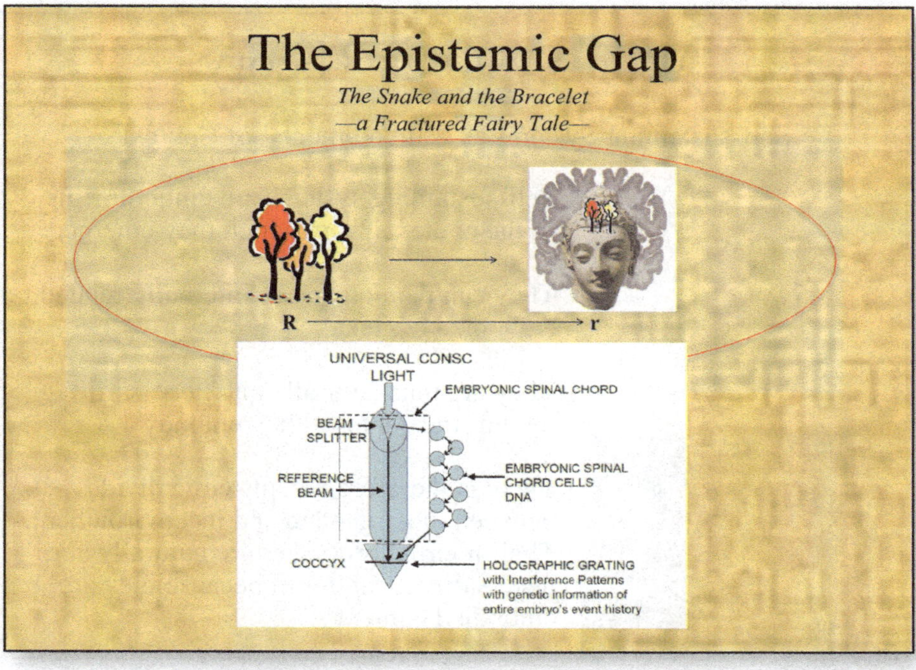

Figure 3

dan and Mays 2017), I ask whether or not it is possible for consciousness to do more. Is the notion of objective-subjective division truly necessary? For Spinoza, reality was the infinite power of Nature, what he also referred to as God. Within this framework, he conceptualized humans as *finitudes*—finite manifestations of infinite Nature, finite manifestations of everything, finite manifestations of God. What we see in Spinoza is a reconfiguration of Descartes' notion of two distinct, interacting kinds (that is, the material and the spiritual) into one (that is, the infinite power of Nature) that we then conceptually divide into two when we try to understand why some aspects of my phenomenology *feel* as if they depend on me (that is, what we label the subjective) while other aspects *feel* as if they do not depend on me (that is, what we refer to as the objective).

WST conceptualizes organisms as *embodiments of context*—a twenty-first century reframing of Spinoza's notion of *finitudes* (**Figure 4**). Instead of partitioning living systems into physical and mental components, WST scientifically describes them as self-sustaining energy transformers that intake, transform, and dissipate energy in ways that sustain the systems of work (that is, energy transformations) that produced them.

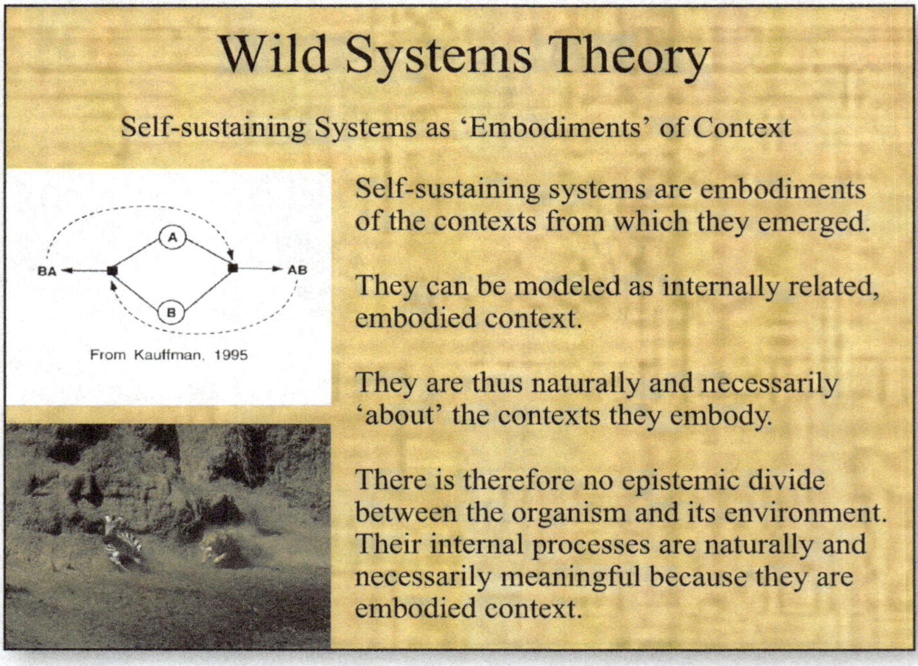

Figure 4

Stuart Kauffman taught us how an energy transformer can sustain its stability in an energy transformation hierarchy. Basically, the idea is that the chemical interactions are the work of the system, giving products that are actually catalysts for the work that produce the catalyst. What you have here is a self-perpetuating, self-sustaining system of work (**Figure 5**). Such systems necessarily give rise to borders that allow them to exist as systems. For a neuron, the border is a bilipid membrane. For a perceiving human, the border is a system-generated, hyper-complex network of neural work that distinguishes changes in neural dynamics due to the system itself from changes produced by the external environment. This self-generated, border-creating, "*other-ing*" is necessarily accomplished by all living systems and gives rise to the *feeling* of there being separate objective and subjective realms. In short, the feeling of system-context separateness is a necessary by-product of the work of *being.*

Given such systems phylogenetically emerged from a specific context (that is, the multi-scale constraints of planet Earth), they necessarily had to embody the constraints of that context. As a result, embodiments of context

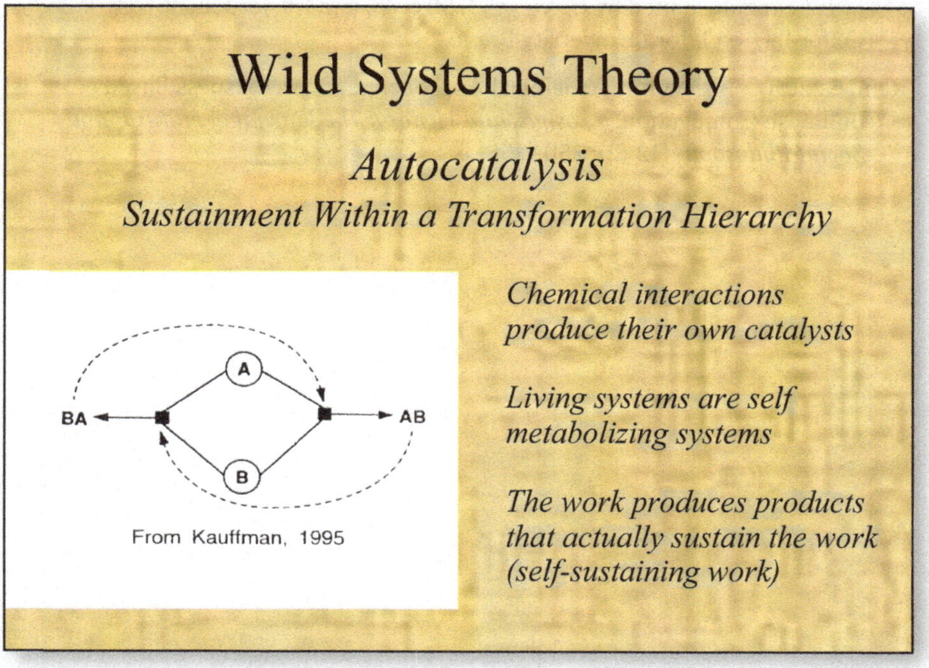

Figure 5

are naturally and necessarily *about* the contexts they embody. The membrane of a single-cell organism is simultaneously *about* the organism as well as the context in which the organism sustains itself. In short, there is no *epistemic gap* between an organism and its context. As finitudes—as embodiments of context—organisms *are* context; organisms *are* aboutness; organisms are meaning. According to WST, then, evolution is about the phylogenetic history of embodied contexts, and science need not be tasked with discovering how physical brains produce subjective aboutness. Aboutness is what we are. ◊

References

Jordan, J.S. 2013. "The Wild Ways of Conscious Will: What We Do, How We Do It, and Why It Has Meaning." *Frontiers in Psychology* 4 (547): 1–12.

———, 2020. "Wild Stories: Science, Consciousness, and the Anticipatory Narratives in which We Live." *Journal of Consciousness Studies* 27 (3–4): 128–151.

———, and C. Mays. 2017. "Wild Meaning: The Intercorporeal Nature of Objects, Bodies, and Words." In *Intercorporeality: Emerging Socialities in Interaction,* edited by C. Meyer, J. Streeck, and J.S. Jordan, 361–378. Oxford, UK: Oxford University Press.

Rein, G. 2016. "A Quantum Chemical Approach to Consciousness Based on Phase Conjugation." *Cosmos and History: The Journal of Natural and Social Philosophy* 12 (2): 250–258.

Session 3 | Second Commentary
On the Presentation by Dr. Glen Rein

How Does Human Consciousness Interact with the Brain?

Dr. Hiroshi Ishii
Hospital Director, Kawasaki Kokoro Hospital, Miyagi, Japan

The mind-brain problem is a very important and difficult one, with a long history that dates back to ancient Greece. There are monistic views, dualistic views, interactionism, idealistic views, materialistic views, and so on. These views strongly influence one's perspective on life and the world, resulting in different lifestyles and cultures. Historically, it has been a problem in the realm of philosophy, but recently it has become a subject of neuroscience. I made a presentation on this topic at the International Symposium on Unification Thought in 2006, and I wrote a book, *The Mind-Brain Problem and Unification Thought* (2008, 2010). My impression is that there are various ideas and theories on this topic, and growing discrepancies among them.

About the Speaker

- Director, Kawasaki Kokoro Hospital, Japan
- Specialization: Geriatric Psychiatry
- Author, *The Mind-Brain Problem and Unification Thought*
- MD, PhD, Geriatric Neuropsychology

Summary of Dr. Rein's Presentation

Dr. Rein takes a dualistic view regarding the mind-brain problem and sees consciousness as the fundamental property of reality. So, the brain and neurons are not the basis of consciousness, but they are the supporting system for consciousness. He thinks that a quantum mechanism is involved in consciousness-brain interactions, especially related to phase conjugation (PC). Optical phase conjugation is a kind of optical wave transmission, which propagates backward toward the source. Nonlinear couplings between incident and reflected waves are made in PC phenomena, producing virtual waves. Porphyrin networks are one of the PC mirrors of the brain, which create quantum holograms.

Here, the mind is defined as a quantum hologram and virtual energy, and PC mechanisms make it possible to receive universal consciousness to individual consciousness. This model also explains how the mind is created as its own hologram right after the brain is formed in the embryo.

What are the new and interesting points of Dr. Rein's model?

Although scientific materialism is prevalent in the modern scientific community, it is only a hypothesis that ultimately has not been proven. Against such materialism, Dr. Rein proposes that consciousness is a fundamental property of reality. I support his viewpoint, and I think we need to develop neuroscientific research from that viewpoint more than ever.

Quantum physics has some mystical aspects. Some scientists call it "weird"; others call it the "quantum enigma." Some researchers think that quantum phenomena are connected to our consciousness directly. Dr. Rein's proposal of a PC mechanism of quantum dynamics is a totally new idea for the mind-brain interaction theory. It would be good to consider virtual and holographic views of the mind when reflecting on the relationship between physical phenomena and mental phenomena. I am interested in the porphyrin network as a possible whole-body mechanism of interactions between the mind and body, because the blood circulatory system carries hemoglobin to every part of the body.

Questions about Dr. Rein's model

What is the relationship between virtual and holographic energy? Are they similar or totally different features? Ordinary neuronal activities are electrical and chemical signal transmissions. Does PC entail those ordinary neuronal activities? Does PC have a relationship with microtubules, as in the model proposed by Roger Penrose and Stuart Hameroff? Does virtual and holographic energy relate with spiritual phenomena? How can we research PC activity of the brain? How can we measure it? Is there any clinical application of PC? How does universal consciousness become individual consciousness? We have five senses of perception: seeing, hearing, smelling, touching, and tasting. Is every sense of perception connected with the PC mirroring system? At which level of the neuronal system is quantum mechanics involved?

Expectations for further research

As mentioned earlier, our view of the mind-brain relationship has a direct influence on our view of life and our culture. I hope that through future research, Dr. Rein will discover the true nature of human beings and the meaning of life. I am a geriatric psychiatrist, and I wonder whether PC or virtual energy has any correlates with psychiatric symptoms like delusions or hallucinations. If so, research in this field might result in a new kind of treatment for psychiatric patients.

The mind-brain problem is also the mind-body problem. If we could know the correct relationship between our mind and body, it would become easier to control ourselves. I agree with Dr. Rein's view, which regards neural systems as supportive of consciousness. I also agree that some quantum mechanisms are related to consciousness and the brain system. I am looking forward to further developments of Dr. Rein's study that would help clarify some mysteries of the mind-brain problem. ◊

Session 3 | Presentation Two

Can Investigations in Frontier Science Lead to the First Cause?

Dr. William Keepin
Cofounder, Satyana Institute, Seattle, USA

There is a secret turning in the heart,
which makes the entire universe turn.
—Rumi

Greetings! It is a great honor, privilege, and I would add *responsibility*—to speak to the topic proposed for this presentation by the organizers of this conference. Let us begin with a brief look into "frontier science," which could be characterized as the scientific exploration of non-material dimensions of reality and consciousness, including phenomena regarded as anomalous or "paranormal" from the mainstream scientific materialist perspective.

About the Speaker

- Cofounder of the Satyana Institute, Seattle, USA, and Inner Pathways Training Centre, Vernon, British Columbia, Canada.
- Cofounder of the Gender Equity and Reconciliation International project
- Organized six international interfaith spirituality conferences. Has studied and practiced intensively in Eastern and Western spiritual disciplines for 35 years–Christian, Hindu, Buddhist, and Sufi traditions
- Author of *Belonging to God* and *Divine Duality* and coauthor of *Song of the Earth*
- PhD in Applied Mathematics

Part I: "Extraordinary claims require extraordinary evidence"

This is the oft-heard refrain—and *demand*—from materialist scientists and skeptics when confronted with extraordinary claims from "frontier science." Today, the demanded extraordinary evidence is *in*. Rigorous, replicated, voluminous evidence is now available in multiple forms, pointing to subtle realities of a spiritual, archetypal, numinous, or metaphysical nature.

A good bit of this evidence has been presented in this conference, and the ensuing revolution is just beginning. Taken as a whole, these new data are ushering in a new, postmaterialist science that integrates the physical universe with the non-physical, "higher" dimensions of reality.

In 1990, the Dalai Lama gave a talk in which he said (paraphrasing):

> *To discover the deepest secrets of the cosmos,*
> *there are four fundamental steps:*
>
> 1. *Explore the outer (visible) universe—physis*
> 2. *Explore the inner (invisible) cosmos—psyche*
> 3. *Explore the relationship between them.*
> 4. *From these three, one can discover the deepest secrets of reality.*

Frontier science provides extensive evidence for step 2 in the Dalai Lama's schema, the science of the nonphysical realms. This extraordinary evidence constitutes a major challenge to the prevailing mechanistic scientific materialist paradigm, which insists that nothing exists beyond the physical universe, corresponding to only step 1 of the above schema. So, the data from frontier science are widely ignored or denied today.

Yet the tables have now turned: extraordinary *denial* of frontier science demands extraordinary *refutation* of the voluminous evidence. The onus is now on the scientific materialist community to either *refute* or *explain* this evidence; and so far, neither is forthcoming. Failing this, the scientific materialist community must either acknowledge this extraordinary evidence as legitimate, including its far-reaching implications, or else find themselves in stark betrayal of the very scientific principles they purport to cherish and uphold.

One of the many areas of frontier science is the field of near-death experience (NDE). The major NDE researchers have concluded that human consciousness survives physical death. This contradicts orthodox scientific materialism, which assumes that life ends with the demise of the body, as if the human being were a kind of biological robot. Yet NDE research and the related field of mediumship research show otherwise, constituting a revolutionary breakthrough.

This breakthrough is also broadly consistent with the spiritual teachings of all the major world religions, which affirm an afterlife in some form for the human soul after physical death. For example, in the *Bhagavad Gita* (a key scripture from India), the first spiritual teaching in Chapter Two is that the human being does *not* die when the body dies. The *Bhagavad Gita* was written *ca.* 500 BCE, so science is just now catching up to this key reality, which religions have known for at least 2,500 years.

Part II: The fractal holoverse

Physicist David Bohm was one of the earliest contemporary scientists to take seriously spiritual realities (Bohm 2002). Bohm was a quantum physicist of the highest caliber and a close colleague of Albert Einstein. Bohm also carried his quest for truth beyond traditional science, engaging in extensive dialogues with spiritual masters such as Krishnamurti, the Dalai Lama, and others.

As Bohm discovered, the major spiritual and religious traditions of the world agree on several key foundations, which they arrived at independently of one another. The religions agree that the architecture of the cosmos consists of multiple ontological levels—ranging from an unseen, ineffable "Absolute" at the highest level, down to the physical plane of matter, space, and time. There are many models of this, but all traditions identify *at least* three fundamental levels of reality: (1) the Absolute (which goes by several names: Godhead, Dharmakaya, Parabrahman, and so on), (2) the "heavenly" planes or archetypal realms, and (3) the physical universe. Many traditions refine this categorization further into seven or ten or more levels.

On each plane of reality, there are definite operative principles and laws, and everything proceeds on each level according to the laws of that level.

Higher planes of reality entail "higher" laws, which often transcend (or supercede or "violate") the laws on lower levels. The principles and laws of each level are just as rigorous and inviolable on their own plane, as physical laws and principles are on the physical plane (such as gravity and the other three fundamental forces). Higher laws sometimes "intervene" or function on lower levels, which is perceived as "miracles" or "magic" on the lower planes of existence. Hence magic is real, as is gnosis, revelation, and so forth—all of which follow "higher" laws, as Dean Radin's work on magic has begun to show in scientific terms (Radin 2018).

Frontier science gives us strong evidence that there are forces and realities—often dubbed "paranormal"—which are actually quite normal in occurrence, but they operate on higher planes of reality. These higher forces organize and create what manifests on lower levels. What is "one" on a higher plane often becomes "many" on a lower plane. These understandings are found in all spiritual traditions, which teach that invisible forces create and organize the visible universe.

David Bohm sought to integrate scientific truth and spiritual truth to form a unified, comprehensive model of reality. He was one of the first contemporary scientists to call for a new science of consciousness. Going beyond energy and matter as the foundational ground of science (culminating in Einstein's mass energy equivalence, $E = mc^2$), Bohm affirmed that there are *three* fundamental constituents of reality: matter, energy, and consciousness (or meaning), each of which enfolds the other two. This culminated in Bohm's theory of the implicate order and the "holomovement"—a concept in which the whole of reality is in constant flux between enfoldment and unfoldment. In this theory, the cosmos is understood to be a single, "undivided wholeness," closely analogous to spiritual cosmologies that speak of universal oneness, such as *advaita vedanta* (Hinduism), or *tawhid* (Islam), and so on. The various ontological levels of reality are embedded as different levels of "implication" in the (unseen) implicate order, and the 'lowest' level of reality is the unfolded "explicate order" (visible universe). In Bohm's words, "Everything is connected to everything else," and this key insight has been recognized in recent years as "quantum entanglement." This term may be something of a misnomer, because the cosmos is not a tangle or chaotic jumble. Perhaps "quantum communion" or "quantum interbeing" (borrowing on Thich Nhat Hanh) would be a better term to express this exquisite oneness, in which all things are interconnected in a seamless, cosmic harmony.

Bohm's work was a major advance, yet sadly, it was largely ignored during his lifetime by the mainstream scientific community. However, it has been receiving increasing attention in recent years, and is summarized in the recent documentary film, *Infinite Potential.* The holographic principle has also been proposed recently as a means to reconcile quantum physics and relativity theory, with the profound implication that the physical universe is a holographic projection of information encoded in a two-dimensional hyperplane. This model also closely corresponds with spiritual understandings of the universe.

Bohm's holomovement proposes that the nature of reality has a holographic or fractal structure. The simplest form of a fractal is the Mandelbrot set, which is a mathematical set in the complex plane, shown in **Figure 1**. It is generated by repeated iteration of the simple formula, $z_{n+1} = z_n^2 + c$.

The Mandelbrot set has a remarkably exquisite structure, which can be seen by zooming in repeatedly, starting with **Figure 1**. If we zoom in right near the top of **Figure 1,** we get **Figure 2**, and if we zoom in again near the top, we get **Figure 3**. If we now zoom into the center of **Figure 3**, and then keep zooming into the center of each subsequent figure, we get the series, shown in **Figures 4** through **8**. Each of these figures is magnified from the previous one by about one order of magnitude.

Figure 1 A Mandelbrot set.

Figure credits: Unless otherwise indicated, all figures were contributed by the author, Dr. William Keepin

Figure 2 Produced by zooming in near the top of **Figure 1**.

Figure 3 Produced by zooming in near the top of **Figure 2.**

Figure 4 Produced by zooming in near the center of **Figure 3**.

Figure 5 Produced by zooming in near the top of **Figure 4**.

Figure 6 Produced by zooming in near the top of **Figure 5**.

Figure 7 Produced by zooming in near the center of **Figure 6**.

Figure 8 Produced by zooming in near the top of **Figure 7**.

Take special note of **Figure 8**. It contains an exact replica of the original set in **Figure 1**, but it is 136 million times smaller. Although very tiny, it has the same structure and all the intricate details of the original. If we were to keep zooming in from here, we would continue to find more embedded miniature replicas of the set, *ad infinitum.* Mandelbrot zooms that continue for many orders of magnitude can be readily found on YouTube.

Figure 9 Ferns Credit: Sanjay ach, CC BY-SA 3.0

Figure 10 Romanesco broccoli Credit: Ivar Leidus, CC BY-SA 4.0

Figure 11 Tree branches
Credit: Łukasz Smolarczyk, CC BY-SA 4.0

Figure 12 Highly branched channels of lightning.
Credit: Griffinstorm, CC BY-SA 4.0

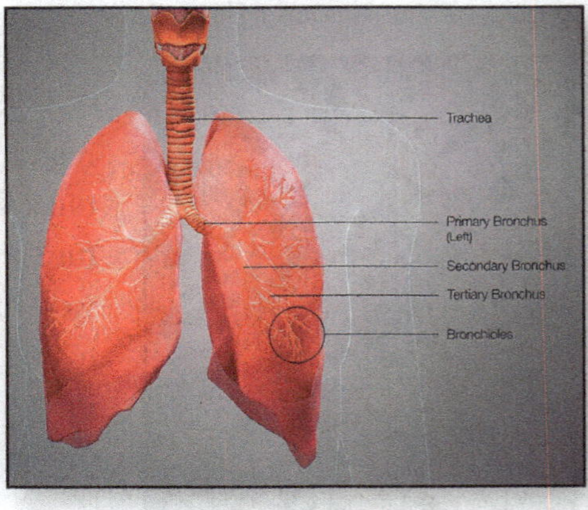

Figure 13 Branching of bronchi in human lungs.
Credit: www.scientificanimations.com, CC BY-SA 4.0

Fractals are not just mathematical entities; fractal structures are found throughout nature. The following are some examples: ferns (**Figure 9**), Romanesco broccoli (**Figure 10**), tree branches (**Figure 11**), lightning (**Figure 12**), and the bronchi in human lungs (**Figure 13**). In each case, these natural fractal creations exhibit "scale invariance": one small part of the structure replicates the superstructure of the whole.

The mathematical process of fractals is schematically illustrated in **Figure 14**. From the "mathematical infinite" of all possible formulas, one "fractal seed" formula was selected, shown in the center, which

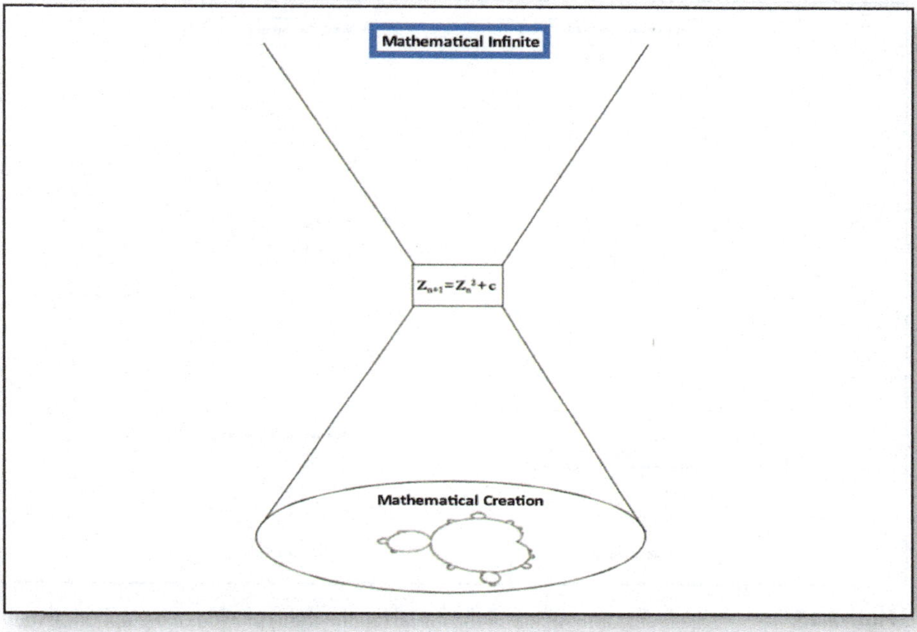

Out of the infinite possible mathematical formulas, a "fractal seed" formula was selected, and it creates a Mandelbrot set. **Figure 14**

creates a Mandelbrot set. The key insight here is that an exceedingly simple formula (or principle) can create an implausibly vast and intricate, complex manifestation. This astonishing superstructure of worlds of infinity within worlds of infinity all comes from one simple, invisible "fractal seed," or source, which creates it all.

Part III: Fractal Consciousness and the "First Cause

I propose that *consciousness itself has a fractal structure.* The mathematical fractal could be considered a mere "footprint" (in the mental domain of mathematics) of a far more general, universal principle of spiritual causation. This helps us understand many aspects of consciousness, including the nature and structure of spiritual traditions.

We see this by analogy in **Figure 15**. On the right is the fractal causation we just saw, and on the left is a fully analogous process, in which a single order-

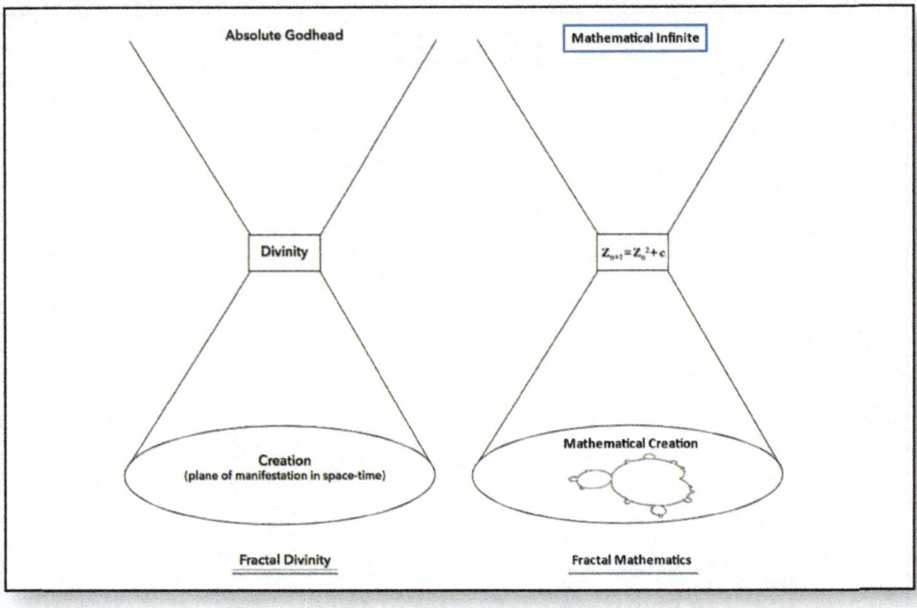

Figure 15 The emergence of the fractal (on the right) provides a metaphorical map of how the creation may have emerged from a divine source (on the left).

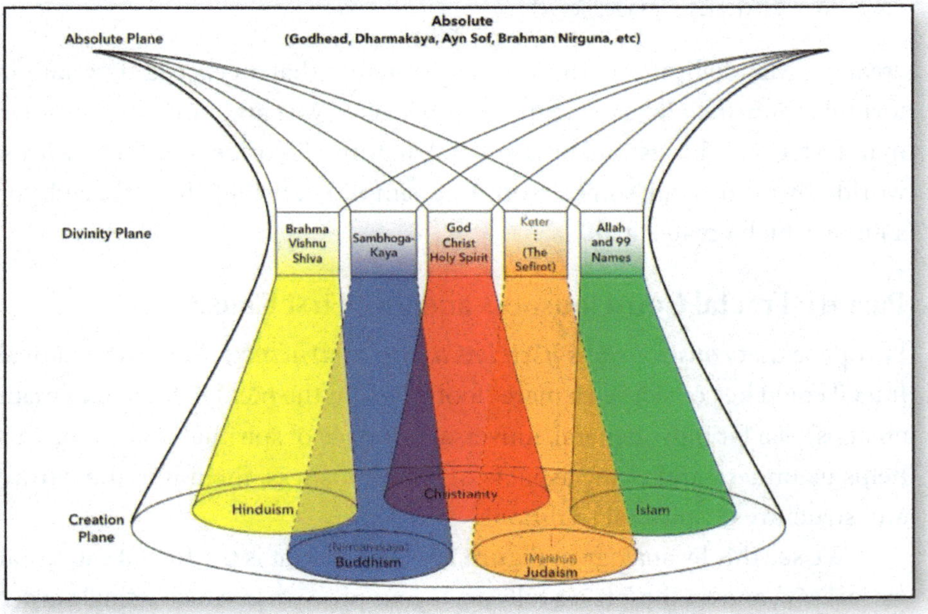

Figure 16 The circular plane at the bottom represents the physical space-time universe that science has been traditionally investigating. The vertical dimension represents various levels of the "implicate order."

ing principle of divinity creates an entire spiritual reality on the plane of creation. The fractal provides a metaphorical map that enables us to envision, by analogy, the possibility that creation itself also emerged from a single, invisible seed of "consciousness," a simple divine source—which several scriptures across the spiritual traditions in fact claim—as the "First Cause" of existence.

Building on these insights, we construct a simple schematic of spiritual reality as a fractal holoverse in **Figure 16**, which integrates the theological and ontological structures of the major religions with the physical universe. The vertical dimension here illustrates the various levels of the implicate order, and the horizontal, circular base at the bottom represents the explicate order: the physical space-time universe, which is the plane of normal science.

Here we see that the religions all have the same ontological superstructure, across the different levels of reality. In each case, the Absolute (which goes by different names in different traditions) emanates a unique divine principle or seed, which in turn produces an entire tradition in the manifest plane. We see a homeomorphic equivalence of the Christian trinity, Hindu trinity, Buddhist sambhogakaya, and so forth, on the plane of divinity.

Across the spiritual traditions, the most advanced masters and mystics generally proclaim identity with the Supreme — including Jesus ("I and the Father are one"), Al Hallaj ("I am the Supreme Reality"), and the Hindu mystics ("I am Brahman") (**Figure 17**). In both theistic and nontheistic traditions, the inmost essence of the human being is none other than the

Oneness with the Supreme source

- **I am Brahman** *Aham Brahmasmi*
  (Upanishads, ***Hinduism***)
- **I and the Father are One**
  (John 10:30, ***Christianity***)
- **I am the Supreme Reality** *Anal haqq!*
  (Al Hallaj, ***Islam***)

Across various spiritual traditions, the most advanced masters and mystics have proclaimed oneness with the Supreme. **Figure 17**

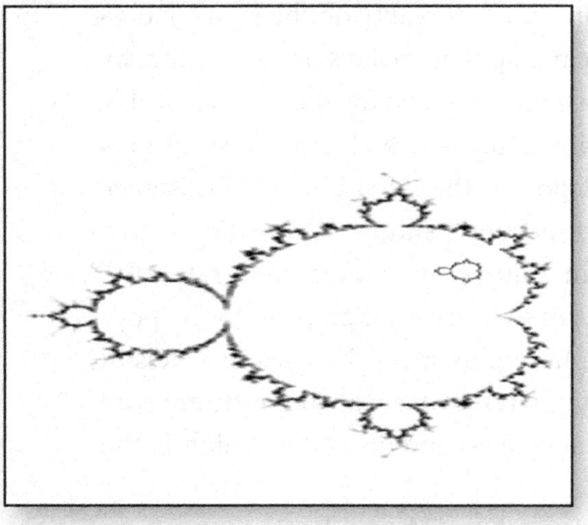

Figure 18 Imagine that the entire cosmos is represented by a Mandelbrot set shown here. The tiny Mandelbrot shape within it represents a single human being who is a spiritual seeker.

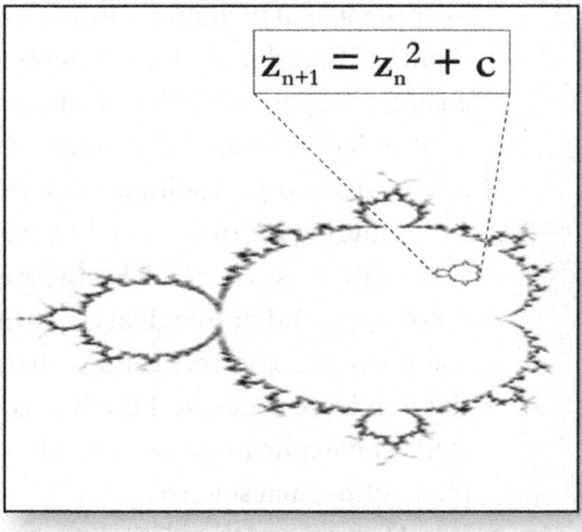

Figure 19 The emergence of a Mandelbrot shape from a mathematical formula could be considered a metaphor for a spiritual seeker who realizes the Source of her own existence.

transcendent essence of the Absolute.

Yet a key question emerges: How can the infinity of divinity possibly fit inside the finite human being? A simple metaphorical thought experiment sheds light here. A schematic of the entire cosmos, represented by the Mandelbrot set, is shown in **Figure 18**. Also shown is a single human being, a spiritual seeker, represented by the tiny Mandelbrot shape in the same figure.

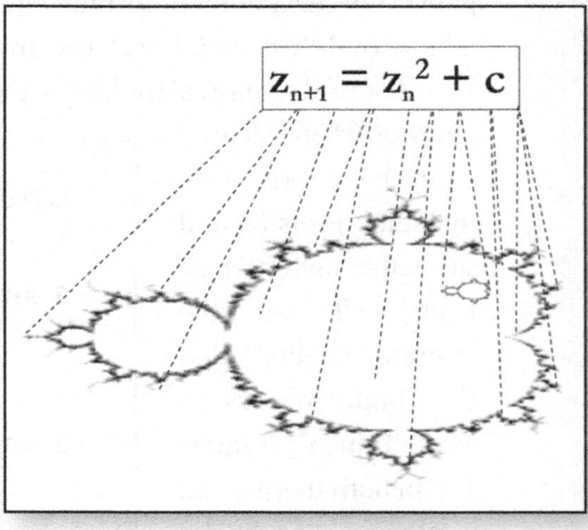

Figure 20 The Source is expressed in all that exists.

Now, through intensive spiritual practice, this spiritual seeker realizes the deep truth and source of her own existence, which is (in this metaphor) the formula, $z_{n+1} = z_n^2 + c$. Her spiritual awakening is illustrated schematically in **Figure 19**.

After that—*voila!*—this seeker has a major spiritual epiphany: her own identity ($z_{n+1} = z_n^2 + c$) is revealed to be none other than the Source of all that exists (**Figure 20**). The seeker realizes, "This, that *I am,* is none other than the very Source of all that is. I *am* the Word, the Seed, the Formula, that gives rise to the entire cosmos." As Rumi expressed this (Barks 2004), "There is a secret turning in the heart, which makes the entire universe turn."

This hidden secret or spiritual essence, deep in *every* heart, is none other than the fractal seed of the entire divine creation. It is spoken of across the scriptures in different ways; in the Gospel of John as the *logos* ("Word") that created everything. It appears in the Bhagavad Gita as the *bijam*, the eternal seed, which is essentially identical (or "homeomorphically equivalent") to the *logos* in Christianity. Notice the remarkable similarity of this foundational source, as proclaimed in scriptures West and East:

> *In the beginning was the* logos *("Word"), and the* logos *was with God, and the* logos *was God. He (*logos*) was with God in the beginning. Through [the* logos*] all things were made; without him [*logos*] nothing was made that has been made.* — Gospel of John 1:1–4

> *I am the original seed (*bijam*) of all existences (7:10). I am the origin of everything, and from Me all proceeds (10:8). I am the seed-giving father (14:4). My eternal seed is to be found in every creature (7:10). Nothing higher than Me exists (7:7).* — Bhagavad Gita

In both cases, the *logos* or *bijam* functions as a fractal divine seed; they can be considered two different names for the same foundational creative principle. In both cases, this source became incarnated in human form: the *logos* became Christ, and the *bijam* became Krishna.

This is the mystic realization of the "First Cause" of creation—something utterly invisible to the instruments of today's science. Yet this realization is not necessarily *un*scientific; we need only expand what is admissible as "scientific data" to encompass the spiritual revelations of the heart. These realizations were independently discovered, and verified, by sages across diverse spiritual traditions and cultures of the world.

Part IV: Archetypal Cosmology: Reconnecting Humanity with the Cosmos

Everything is connected to everything else; thus says quantum entanglement (or "quantum interbeing"). This extends to interconnections between physical and consciousness domains. Seemingly disparate material objects and consciousness events—which appear totally unrelated and *dis*connected to our physical senses and scientific instruments—are actually deeply *inter*connected on higher levels. This holds because quantum entanglement functions not only "horizontally" throughout the plane of the physical universe, but also "vertically," unifying multiple levels of reality. This phenomenon was termed "acausal synchronicity" in the pioneering collaborative work of Carl Jung and Wolfgang Pauli (Jung and Pauli 2012).

The synchronistic interconnection between consciousness and cosmos finds surprising new scientific evidence in the controversial domain of correspondence between celestial and human events—better known as the controversial discipline of astrology. This and other esoteric sciences (such as alchemy, gnosis, Tarot, and *I Ching*) are generally dismissed as "pseudosciences" worthy of ridicule. Yet the shocking truth is that some of these disciplines actually provide remarkably strong *scientific* evidence for close interconnections between the visible and invisible domains of reality, thereby providing substantive evidence for Step 3 of the Dalai Lama's schema. Following Gary Schwartz's example earlier in this conference, I am taking the unusual step here of sharing some of this controversial yet little-known research.

About 70 years ago, French statisticians Michel and Françoise Gauquelin set out to disprove astrology once and for all, using rigorous statistical analysis. They generated a comprehensive database of some 23,000 human subjects, and indeed found no evidence for certain astrological claims. However, to their complete shock, they discovered several statistically significant correlations linking the positions of specific planets (relative to the Earth) at the birth times

Planetary correlations with eminent leaders ("Mars effect")

- **Sports champions** (n = 2,144), **Mars**
- **Scientists** (n = 1,094), **Saturn**
- **Writers** (n = 1,352), **Moon**
- **Actors** (n = 1,407), **Jupiter**
- **Journalists** (n = 673), **Jupiter**
- **Military** (n = 3,608), **Mars, Jupiter**
- **Physicians** (n = 2,552), **Saturn, Mars**
- **Politicians** (n = 1,002), **Jupiter, Moon**

Composite probability by chance: p < 1 in 5 million
Independently replicated at least 14 times

Figure 21

of eminent leaders in various fields: Mars for sports champions, Saturn for scientists, Jupiter for actors and politicians, and the Moon for writers (**Figure 21**). Their data exhibited a relatively weak effect, yet highly significant with a probability of occurring by chance of less than 1 in 5 million.

These results were understandably received with tremendous skepticism, and several scientific groups set out to debunk them, or discover the presumed artifact responsible for them. Yet in several carefully controlled independent studies, would-be debunkers, to their dismay, replicated the results. The Gauquelin findings have now been replicated at least 14 times. The Gauquelin database has also been further analyzed in numerous meticulous studies, using far more sophisticated multi-factor methods which were not available to the Gauquelins at the time, and have revealed many *additional* astrological correlations in their data of strong statistical significance.

Scientific tests of astrology have meanwhile become far more sophisticated in recent years, applying multi-factor methods, big data analyses, and large sample sizes. Today, over 70 scientific studies published in peer-reviewed journals support astrology (Currey 2020). Most (but not all) components of traditional astrology have received strong statistical validation (Tarvainen 2021). Of course, correlation does not imply causation, and these findings remain unexplained theoretically.

These recent studies overturn the majority of earlier scientific studies of astrology, including the most famous study published in *Nature* by Shawn Carlson in 1985 (Carlson 1985). Carlson's double-blind study reported that astrologers failed to match the astrological natal charts of 116 human subjects to their respective psychological profiles at better than chance levels. Re-analysis of this study in 2009 uncovered substantial errors that, when corrected, reversed the results to demonstrate modest statistical support favoring astrology (Ertel 2009; Currey 2011). Nevertheless, the Carlson study is still widely though erroneously cited as the "definitive" scientific study debunking astrology (Stierwalt 2020).

Recent research dramatically supercedes the Carlson study. An astrological expert system was developed by compiling key words (drawn from 25 leading astrological texts) for more than 5,000 specific astrological factors. This expert system software simulates a full astrological chart reading, essentially 'replacing' a human astrologer (thereby eliminating human subjective bias, intuitition, psychic perception, or 'cold reading'). This expert system was applied in automated blind-matching studies conducted with 77 celebrities, and it matched astrological horoscope charts with their owners—using key words from published biographies—at p-values (probabilities by chance) ranging between 5×10^{-4} and 1.2×10^{-9}. Because "human intervention was eliminated from the match-making, and the evidence far exceeds chance, this study suggests that there is an objective basis for astrology" (Godbout 2020, 2021).

The statistical evidence from this and other recent astrological studies is sufficiently precise that it is beginning to be applied directly in clinical practice of astrology (Tarvainen 2021). Other recent big-data studies enable formal calculation and optimization of various astrological parameters, which have enabled independent calculation of certain ancient astrological protocols (such as planetary orbs). Recent studies also provide statistical confirmation of specific practices employed by Ptolemy and Vedic astrologers going back more than 2,000 years. This appears to be another arena where contemporary science is just now discovering hidden truths well-known to ancient sages.

Of course, there are many scientific studies that have found no astrological effects. Most of these studies are either substantially flawed or employ data sets that are too small to detect the effects. Either way, as the saying goes, it takes only one black swan to prove that not all swans are white. Therefore, the onus is now on the scientific skeptics to either *refute* or *explain* the 70 "black swan" studies that affirm astrology. So far, such refutation is not forthcoming, and these recent advances in astrological research stand as a profound and potentially paradigm-breaking challenge to contemporary scientific understanding of the nature of the planets and the cosmos, and their possible interconnection with human consciousness and terrestrial events. Nevertheless, this body of pioneering research remains largely ignored in the scientific mainstream.

In addition to these scientific studies, extensive clinical research conducted by psychiatrist Stanislav Grof, Richard Tarnas, and others has shown that certain astrological indicators, known as planetary transits, provide an extremely practical tool for accurate *prediction* of the precise timing and archetypal character of their clients' psychological experiences in deep experiential psychotherapy and in psychedelic psychotherapy sessions. These researchers found that different types of psychological experience are closely correlated with specific positions and movements of the planets Saturn, Uranus, Neptune, and Pluto, in relation to the natal astrological charts of their clients. Difficult as it may be for some to believe, it is a fact that these astrological transits have proven to be far more accurate and valuable in clinical practice of experiential psychotherapy than any of the standard psychological and personality profile tests, which these researchers also investigated and applied systematically, without success or useful results for these purposes.

In summary, as psychologist Hans Eysenck wrote, "Perhaps the time has come to state quite unequivocally that a new science is in the process of being born" (Eysenck and Nias 1982). This could precipitate a new scientific revolution, in our understanding of the cosmos and its relationship to humanity.

How could the planets or stars possibly have anything whatsoever to do with human affairs? The idea seems utterly preposterous! Yet, a plausible hint comes from the ancient vision of *Indra's Net* in Buddhist and Hindu cosmology, which illustrates the fractal unity of spiritual consciousness. Reality is imaged as a vast (perhaps infinite) lattice of magnificent jewels, each of which reflects all the others. If you look into the facets of any one jewel, you see all the others reflected there. If we bring this exquisite image 'down' into space-time manifestation, this introduces spatial and temporal variation: the jewels are now of all different sizes, whirling around in grand dynamic motion.

If you now gaze into the facets of any one jewel, you will see all the movements of this entire "universe" of swirling jewels, precisely replicated in dynamic miniature. Each jewel thus "contains" the dynamic fractal microcosm of the entire cosmic process, as it unfolds in holographic evolution on multiple levels of space, time, and consciousness simultaneously. This gives rise to the observed correlations across celestial and terrestrial domains and scales, all the way down to the individual human being, who is evidently a walking miniature cosmos. Perhaps this is one of the secrets revealed in step 4 of the Dalai Lama's schema.

In short: Take Indra's Net, dip it in space-time, and you necessarily get an astrological hermeneutic. This may also be the deeper implication and practical application of David Bohm's holomovement, which holographically interweaves matter, energy, and consciousness into an intricate, undivided, fractal wholeness at every level and scale of existence.

These preliminary theoretical speculations are admittedly pure imagination at this point. Nevertheless, as Einstein said (Einstein 2009), "Imagination is more important than knowledge. For knowledge is limited, whereas imagination embraces the entire world, stimulating progress, giving birth to evolution. It is, strictly speaking, a real factor in scientific research."

Part V: Conclusion: Science Needs an Epistemology of the Heart

In conclusion, a **two-fold scientific revolution** is needed today. On the one hand, we need a **"frontier science" revolution,** and especially the development of a new science of consciousness. This is underway to some extent, but it is still actively resisted by the scientific materialist mainstream and needs to be stepped up. This is a "normal" scientific revolution, precipitated by the "extraordinary evidence" from frontier science, and necessitated by the imperatives of scientific integrity to follow where the data and evidence lead. This will require the entire current materialist, mechanistic scientific paradigm to be replaced by a broader paradigm that is more accurate, subtle, and comprehensive—and thus transcends yet still includes the older scientific paradigm as something of a limiting case.

On the other hand, we also need an unprecedented, **epistemological scientific revolution.** This is not a "normal" scientific revolution, because this one will redefine the very nature of scientific epistemology itself, and significantly expand what constitutes legitimate scientific methodology and data. This entails a kind of Copernican revolution of epistemology. Just as the Earth was once seen as the center of the universe, around which the entire universe revolved; likewise, in today's science, the conceptual mind is viewed as the foundation of scientific epistemology, around which all scientific truth and reality are constellated. This revolution acknowledges the inherent weaknesses, biases, and blind spots of the conceptual mind and its way of knowing, and it reaches beyond the mind to develop a new **epistemology of the heart.**

The "heart" is the primary organ of spiritual perception in the human being. It unifies all the various ontological levels of reality. If science is ever to evolve and open to the deep truths of higher levels of reality, it needs to develop its own scientific "epistemology of the heart." This transcends yet includes the domain of the mind.

I readily acknowledge all the delusion, abuse, and exploitation that has taken place in the name of religion and spirituality. Nevertheless, it is also true that within the religious and spiritual traditions can be found true mastery of the inner journey of the heart, which unveils the spiritual depths of the human being. We suffer today from a case of mistaken identity. The epistemology of the heart is a key element of spiritual practice that can lead us to correct this mistaken identity.

As Plotinus said, "the soul suffers when the mind resorts to evidence and proof." The mind's insistence to explain and understand everything on its terms is a major limitation that is holding science back. As the great Christian mystic St. John of the Cross said, when it comes to the higher dimensions of reality, "We advance by *not* understanding." Or as a Sufi expression puts it, "True understanding is understanding the *in*ability to understand." There is a profound need to develop an epistemology that is rooted in this humility, and it invites a revelatory knowledge, which thereby opens to possibilities beyond what can be conceived or understood by the mind.

In closing, let us return to the opening question: **Can Investigations at the Frontiers of Science Lead to the First Cause of the Universe?** Frontier science has opened the door to invisible dimensions of reality, and provides compelling evidence for a new, postmaterialist science, which could one day lead to scientific discovery of an invisible First Cause of the universe. Frontier science provides remarkable evidence for nonphysical forces and factors that orchestrate the physical universe, which, when taken together, point toward a First Cause, but do not constitute incontrovertible proof. Proof is rare in all science except mathematics. However, there is plenty of evidence, and the real question is, will this evidence be taken seriously?

Science today is in crisis, not just the "replication crisis" of metascience, which is serious enough, but a far more profound crisis. For me, these two scientific revolutions point the way forward for a bright, auspicious future for science. But the opposing forces are formidable, operating in high places with tremendous control. The scientific materialist agenda is to limit the entire cosmos to physical reality alone, collapsing the Dalai Lama's schema to step 1 only.

David Bohm was one of the most brilliant, advanced physicists of the twentieth century, yet even his esteemed colleague and former mentor, Robert Oppenheimer, helped orchestrate a systematic rejection of his views.

Oppenheimer enjoined his peers of the most prominent physicists in the world: "If we cannot prove David Bohm wrong, then we must all agree to ignore him" (Becker 2018). This astonishingly unscientific attitude seems to be the same posture that much of today's mainstream scientific community is taking with respect to the data and evidence presented from frontier science. Nevertheless, there are hopeful initiatives on the horizon, such as the Academy for the Advancement of Postmaterialist Sciences and the Galileo Commission.

In closing, it is a supreme irony that our civilization, steeped in the love of science, has failed to develop a genuine science of love. Ultimately, developing this epistemology of heart is effectively an epistemology of love. It means to embrace and open to the power of love, which is the greatest power in the universe. The cosmos was created as an act *of* love, *by* love, and *for* love. This is a deep truth, which today's science must live into, if it is to serve the future of humanity. ◇

References

Barks, Coleman. 2004. *The Essential Rumi: New Expanded Edition.* New York: Harper One.

Becker, Adam. 2018. *What Is Real? The Unfinished Quest for the Meaning of Quantum Physics.* New York: Basic Books.

Bohm, D. 2002. *Wholeness and the Implicate Order.* New York, NY: Routledge. In addition, the movie *Infinite Potential* is highly recommended.

Carlson, S. 1985. "A Double-blind Test of Astrology." *Nature* 318: 419–425.

Currey, R. 2011. "U-turn in Carlson's Astrology Test." *Correlation* 27 (2): 7–33.

———. 2020. "The Evidence List: 63 Studies." *Correlation* 32 (2). Seven additional studies include five in *Correlation* 33 (2), 2021, and two studies by Renay Oshop (ayurastro.com).

Einstein, A. 2009. *Einstein on Cosmic Religion and Other Opinions and Aphorisms.* Mineola, New York: Dover Publication. 97. Republished from: *Cosmic Religion and Other Opinions and Aphorisms.* 1931. New York: Covici-Friede. Quoteinvestigator.com/2013/01/01/Einstein-imagination/. Accessed May 13, 2022.

Ertel, S. 2009. "Appraisal of Shawn Carlson's Renowned Astrology Tests." *Journal of Scientific Exploration* 23 (2): 125–137.

Eysenck, H., and D. Nias. 1982. *Astrology: Science or Superstition?* London, UK: Pelican Books.

Godbout, V. 2020. "An Automated Matching Test: Comparing Astrological Charts with Biographies." *Correlation* 32 (2): 13–41.

———. 2021. "To use or not use midpoints?" *Correlation* 33 (2): 35–46.

Jung, C., and W. Pauli. 2012. *The Interpretation of Nature and the Psyche* (orig. 1952). New York and Tokyo: Ishi Press.

Radin, D. 2018. *Real Magic: Ancient Wisdom, Modern Science, and a Guide to the Secret Power of the Universe.* New York, NY: Harmony Books.

Stierwalt, S. 2020. "Is Astrology Real? Here's What Science Says." *Scientific American.* www.scientificamerican.com/article/is-astrology-real-heres-what-science-says/. Accessed May 13, 2022.

Tarvainen, K. 2021. "Statistical Studies Have Started to Advance Astrological Techniques." *Correlation* 33 (2): 55–64.

Session 3 | First Commentary
On the Presentation by Dr. William Keepin

Can Investigations in Frontier Science Lead to the First Cause?

Professor Gary E. Schwartz*
Department of Psychology, Medicine, Neurology, Psychiatry, and Surgery
The University of Arizona, Tucson, USA

Dr. William Keepin's presentation was on the topic, "Can Investigations in Frontier Science Lead to the First Cause?" It raises a suite of important and wise questions. It is a privilege for me to provide an invited commentary on his visionary analysis, and I thank the organizers for this opportunity.

Keepin's presentation is organized in five parts. For clarity of discussion as well as honoring limitations in space and time, I will briefly highlight some key points in Parts I–III and then underscore Part V.

Part I

In Part I, Keepin considers a well-known statement made by Dr. Carl Sagan, the distinguished astronomer, public science communicator, and skeptic. Sagan said, "Extraordinary claims require extraordinary evidence." I deeply resonate with the ethical position that Sagan advanced and Keepin has extended.

For example, I published a chapter, "Extraordinary Claims Require Extraordinary Evidence: The Case for Postmaterialist Consciousness," in a book titled, *Expanding Science*, published by the Academy for the Advancement of Post-

* Please see Dr. Schwartz's short biography with his keynote address on page 29.

materialist Sciences (Schwartz 2020). Furthermore, I have completed a book titled, *Extraordinary Claims Require Extraordinary Evidence: The Science and Ethics of Truth Seeking and Truth Abuse* (Schwartz 2021).

I have proposed that extraordinary claims not only require extraordinary *evidence,* but that extraordinary claims require extraordinary *evaluations of the evidence* as well.

Moreover, I have proposed that, for practical as well as ethical reasons, the need to make extraordinary evaluations applies not only to those who make the claims. This need applies to all of us, since we must make decisions about the accuracy of information as we navigate our personal and professional lives.

Importantly, Keepin has rephrased Sagan's statement to read, "Extraordinary claims *demand* extraordinary evidence." This is a powerful restatement of Sagan's position, and it deserves to be highlighted. Moreover, Keepin applies this restatement to pseudo or denial skepticism when he says, "Extraordinary *denial* of frontier science demands extraordinary *refutation* of the voluminous evidence."

Keepin's take-home message is so important that I will quote the remainder of his paragraph here:

> *The onus is now on the scientific materialist community to either refute or explain this evidence; and so far, neither is forthcoming. Failing this, the scientific materialist community must either acknowledge this extraordinary evidence as legitimate, including its far-reaching implications, or else find themselves in stark betrayal of the very scientific principles they purport to cherish and uphold.*

Part II

In Part II, Keepin focuses on a discussion of the fractal holoverse. He honors the brilliant work of Dr. David Bohm and integrates it with the fractal mathematics of the Mandelbrot set (Bohm 2002). Keepin's conclusion is best stated in his own words:

> *The key insight here is that an exceedingly simple formula (or principle) can create an implausibly vast and intricate, complex manifestation. This astonishing superstructure of worlds of infinity within worlds of infinity all comes from one simple, invisible "fractal seed," or source, which creates it all.*

Part III

Extending this vision, in Part III Keepin raises the question of discerning the "First Cause." He proposes:

> *The mathematical fractal could be considered a mere "footprint" (in the mental domain of mathematics) of a far more general, universal principle of spiritual causation.... The fractal provides a metaphorical map that enables us to envision, by analogy, the possibility that creation itself also emerged from a single, invisible seed of "consciousness," a simple divine source—which several scriptures across the spiritual traditions in fact claim—as the "First Cause" of existence.*

In keeping with Keepin's vision, we can envision that Bohm's concept of the implicate order can be expanded to embrace the integrative idea of an implicate, infinitely intelligent, holographic mathematical potential, where fractal mathematics is an illustrative "metaphorical map" of a "mere footprint" of this grand creative potential. Sadly, space and time preclude my reviewing Keepin's beautiful and concise presentation of core historical religious ideas that precede (and, we can say, pre-**"seed"**) these contemporary advances in understanding.

Part V

In Part V, Keepin boldly states that science needs an **epistemology of the heart.** Heproposes that what we need now is a **two-fold scientific revolution** addressing:

1. a **"frontier science" revolution,** in combination with
2. an **epistemological scientific revolution.**

The epistemological side includes recognizing the fundamental importance of mind and consciousness, not only in the existence of everything but in our coming to understand and appreciate it as well. Keepin emphasizes that, "It is a supreme irony that our civilization, steeped in the love of science, has failed to develop a genuine science of love."

In my book, *The G.O.D. Experiments: How Science is Discovering God in Everything, Including Us* (Schwartz 2006), I included a chapter titled, "Infinite Love: The Ultimate Gift from God." This chapter explicates and celebrates the extraordinary capability of humans to learn to love virtually everything in the universe, from the micro to the macro, from the visible to the invisible, from the material to the nonmaterial, and beyond.

Conclusion

Thanks to the organizers of this conference who thought to invite Keepin to give this visionary presentation, we all have the potential to reflect on these grand goals and contribute to their advancement in our own ways, in both our professional and personal lives. Moreover, collectively we can advance the restoration of our species and planet for the benefit of all creation. ◊

References

Bohm, D. 2002. *Wholeness and the Implicate Order.* New York, NY: Routledge.

Schwartz, G.E. 2006. *The G.O.D. Experiments: How Science Is Discovering God in Everything, Including Us.* New York, NY: Atria / Simon & Schuster.

———, 2020. "Extraordinary Claims Require Extraordinary Evidence: The Case for Postmaterial Consciousness." In *Expanding Science,* edited by M. Beauregard, G.E. Schwartz, N.L. Dyer, and M. Woollacott. Battle Ground, WA: AAPS Press.

———, 2021. *Extraordinary Claims Require Extraordinary Evidence: The Science and Ethics of Truth Seeking and Truth Abuse.* Cardiff, CA: Waterside Productions.

Session 3 | Second Commentary
On the Presentation by Dr. William Keepin

Intelligent Design in Planetary Science

Dr. Takahiro Hiroi
Senior Research Scientist, Brown University, Providence, Rhode Island, USA

Thank you, Dr. Keepin, for your insightful presentation. As an example of the need for a paradigm shift from materialist to postmaterialist science, I would like to share the theory of intelligent design based on my field of specialty, planetary science, to derive the existence and nature of God.

Significance of having one large Moon

After Apollo 11 landed on the Moon and returned lunar samples, we discovered that the average oxygen isotope composition of lunar materials is identical to that of terrestrial materials. This finding allowed scientists to nar-

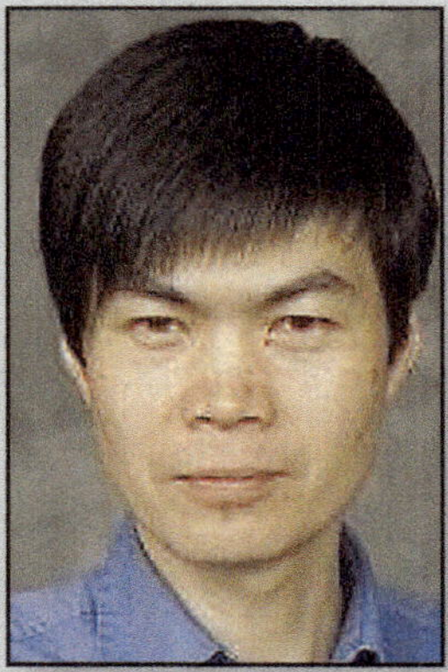

About the Speaker

- Senior Research Associate, Department of Earth, Environmental and Planetary Sciences, Brown University, Providence, Rhode Island, USA
- World Research Institute for Science and Technology
- Planetary scientist specializing in spectroscopy of meteorites and asteroids
- PhD in Mineralogy, The University of Tokyo

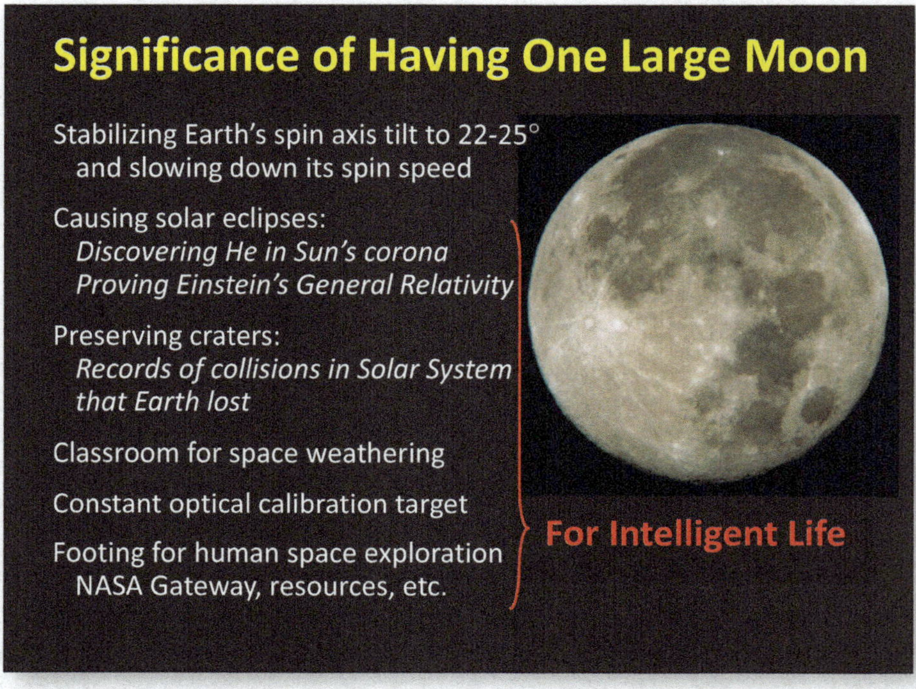

Figure 1

row down the models for the origin of the Moon to the giant impact hypothesis that a Mars-sized body collided obliquely with early Earth to produce mixed ejecta to form the Moon.

Apollo 11 also left a laser ranging retroreflector on the Moon to measure the precise distance between Earth and Moon, which is currently widening by 3.8 cm per year. The Moon used to be much closer to Earth in the past, but it kept moving away from the Earth by the tidal effect that transferred Earth's rotational momentum to the Moon while slowing down Earth's spin.

The existence of one large Moon also kept Earth's spin axis inclination highly stable, at 22–25 degrees, over the entire Earth-Moon history (**Figure 1**). This fact, along with the existence of water and an atmosphere, kept the temperature distribution on Earth more uniform, making its environment ideal for life over billions of years.

Unique characteristics of Earth

Land and ocean: Earth has both land and ocean, which is a very unique combination of features among known planets. These features can be indispensable for intelligent (science-capable) life to emerge. Without an ocean, life would not have been born at all; and without land, humankind would not have emerged.

Magnetic field: Earth's magnetic field is generated by molten iron spinning in the outer core, which protects life from harmful solar wind (energetic positive ions). To generate this strong magnetic field, the Earth needs certain amounts of iron and radioactive materials as heat sources.

Plate tectonics: Earth is the only planet known to have active plate tectonics. This feature allows for the supply and circulation of necessary elements such as hydrogen, carbon, oxygen, and nitrogen to develop and sustain the atmosphere, ocean, and life.

Earth-Moon system as an ideal place for scientific exploration

Solar System in the Milky Way: Earth's atmosphere is transparent mostly over the visible and near-infrared wavelength range, allowing us to observe celestial bodies. The Solar System is currently located between two arms of a spiral galaxy called the Milky Way, and the Earth revolves around the Sun on a plane that is significantly inclined from to the galactic plane. As a result, we are able to direct our telescopes in a dark sky in various orientations to study the galactic structure and farthest stars born in the early universe.

Solar eclipse: The Moon has been moving away from Earth over the last 4.5 billion years, and its current apparent size coincides with that of the Sun. When the Sun was darkened by a solar eclipse, astronomers could observe the light of a star that was on the other side of the Sun. In this manner, they proved a prediction of Einstein's theory of general relativity, namely, that light can bend by gravity.

Lunar surface: Numerous craters on the Moon record the history of past impact events not only on the Moon but also on Earth where they are largely lost. The lunar surface is an ideal classroom of space weathering that chemically or structurally alters materials by the solar wind, radiation, and micrometeorite bombardments.

Space Exploration: The Moon is also a promising place for the expansion of human exploration into space, and as a source of necessary materials such as water (ice) and oxygen, and nuclear energy through helium-3. In addition, the Earth-Moon orbit can be used as a relay station (such as NASA's Gateway) for transportation between Earth, Moon, Mars, and beyond.

Antarctica as a time capsule of Earth and the Solar System

The history of the Solar System has been greatly revealed by studying meteorites, which include fragments of left-over planetesimals, the building blocks of planets. Tens of thousands of meteorites have been found on Ant-

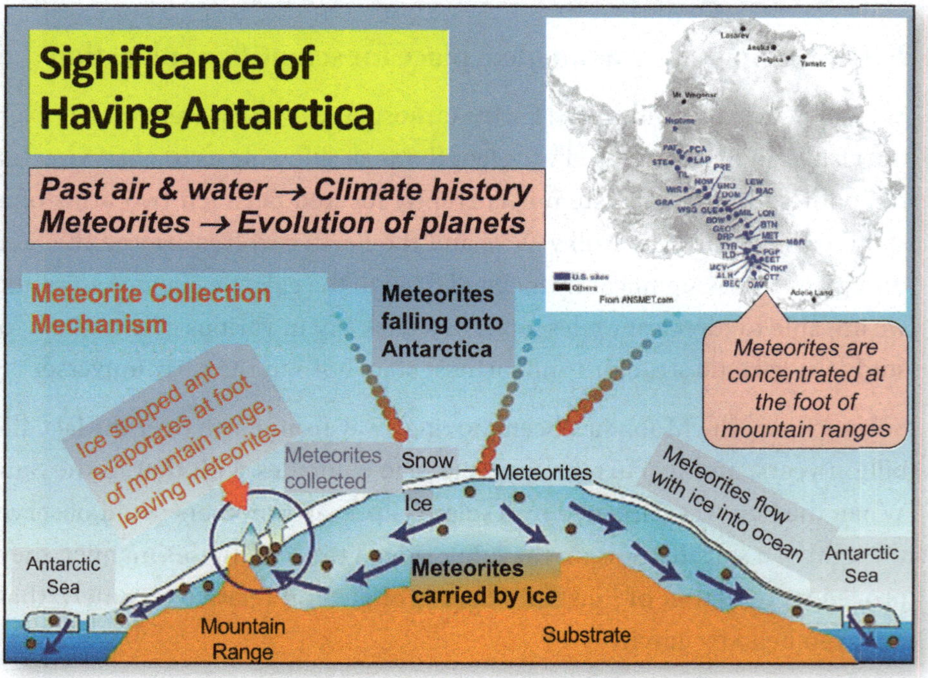

Figure 2

arctica. Those meteorites fell and sank into Antarctic ice, remained frozen, moved with the ice, and accumulated at the feet of mountains, where surface ice evaporated by the heat and light of the Sun (**Figure 2**).

Antarctica also retained ancient air and water, allowing us to derive the history of past atmospheric temperatures. However, Antarctica was not always in its current position on the planet. It moved from the north pole to the south pole and formed its icy layer only within the last 50 million years or so, and it may move away from there in the future. Interestingly, humankind emerged and conquered Antarctica at the right time.

Summary

Figure 3 gives a list of some of the nearly miraculous events in the history of the Earth-Moon system. It strongly indicates that God led every stage of evolution of this planet and its inhabitants, and that He wants us to learn about His creation process through science. ◇

Necessary conditions for mankind to emerge and develop science:

1. *Position & orientation of Solar System in Milky Way Galaxy*
2. *Position of Earth in Solar System*
3. *Angle and speed of Moon-forming giant impact*
4. *Amounts of H, C, O, Fe, radioisotopes, etc. in Earth → Transparent atmosphere, magnetic field, etc.*
5. *Plate tectonics to supply necessary elements to the surface of Earth and form Antarctica at a specific timing*

➡ *God controlled every step of the formation and evolution of Earth and its inhabitants so that humankind could emerge and develop science. He wants us to discover His creation process.*

Figure 3

Session 3

Discussion

Menas Kafatos (Session Chair): I thank the speakers and commentators. Now we are going to have the last part of Session 3: the discussion, with questions and answers. To participate, either raise your hand or put a question in the chat box for one of the speakers. This is the opportunity for the audience to participate and interact with the speakers and commentators. Now we have a question from Dr. Takashi Kato.

Takashi Kato: Thank you very much for the interesting presentation, Dr. Keepin. You quoted many interesting sayings from various religions from earlier ages until now. Do they have an impact on each other? Do you think these traditions must be unified in the future, or do you think the originality of each one should be maintained?

William Keepin: Thank you for that question. My answer is a certain "Yes and Yes!" They should maintain their originality because each one is an utterly unique revelation, an irreplaceable emanation from the plane of the absolute into the manifest plane. At the same time, in all of the major religions, as I understand them, and consistent with what is called the "perennial philosophy" that Dean Radin referred to earlier, there are certain universal truths that they all reveal. Many of the mystics from the various traditions give a very similar account of the highest states of spiritual realization. So the ultimate truth is *one,* and each religion or spiritual tradition is a unique revelation of that truth. Each religion is a sacred world unto itself, and provides a whole *and* complete pathway for the human soul to reclaim its full identity with the absolute—which, as Menas Kafatos said earlier, is found across the traditions. Common to all of them is this Truth, that the inmost essence of the human being is none other than the transcendent essence of the absolute, whether that is called God, or Dharmakaya, or the Tao that has no name, or the Great Spirit. This universal teaching is found across the traditions, and each of the traditions is a particular pathway to realize this truth. So, in summary, the way I see it, the religions are like the *uncreated* Divine Light emanating into *created* light rays of different colors, rather like a refraction into a spectrum of many beautiful colors. The created rays are each unique, yet all are made of light, and their purpose is to lead us back to the Divine Light, the Source itself. They come down to us in these different, beautiful varieties, and these should be retained and respected for their diversity. And because divine consciousness is a fractal holoverse, each of these traditions contains within itself all the others.

Kafatos: The next question is from Richard Lewis.

Richard Lewis: This is for Dr. Rein, who talked very much about neurons. I wonder if you have anything to say about the role of the glial cells, which are predominant in the human brain and make it very unique? Some people have speculated that the glia and the neurons are rather like nucleic acids and proteins—the glia process information, while the neurons do all the analog manipulation of stuff. Do you have any thoughts on that? Then, for Dr. Keepin, I wonder if you have any thoughts about the "humble science" that recognizes but does not know what 75% of the universe is composed of—this dark energy—whether you have any thoughts about how that relates to the spiritual side of things.

Glen Rein: That is a great question. My model has not yet been expanded to consider glial cells, but based on your question, I certainly will do that. I could speculate at this point that glial cells are very likely to be involved. Although they are traditionally believed to have a supportive role, there are other theories that would speculate that the glial cells and the neurons work side by side, as you said, processing the information. Also, the glial cells are believed to function at the electric field level, rather like the DC electric current, rather than the AC system, which is in the neurons. Thus, they complement each other. In that sense, they are two aspects of the same thing. If you consider the brain as electromagnetic in nature, there are two fundamental energies involved: DC and AC. I had not considered that even as a type of energy in the brain, but obviously it is, and I thank you very much for that question. I will try to incorporate that into the next version of my talk.

Kafatos: Thank you. The next question is from Jin Choon Kim. Please tell us whom you would like to address the question to.

Jin Choon Kim: I was so inspired by Dr. Glen Rein and Dr. William Keepin. I have two questions. The first is for Dr. Glen Rein. In your view of quantum biology, you mentioned the existence of a "quantum information field" or "quantum information wave" in all beings. In this sense, brain cells should have a quantum information field. So, is it possible to explain if the quantum information field can connect between consciousness and the brain? Of course, I think our consciousness is composed of a higher mind and a lower mind—or, in another sense, the spirit mind and the physical mind. That is my first question.

My second question is for Dr. William Keepin. You mentioned the "quantum communion." That concept is very interesting. It could be a connector or bridge between spiritual energy and the spirit world, and also between physical energy and the physical world. So, maybe physical energy is like physical zero-point energy, as Glen Rein mentioned, and he may say that spiritual energy is like spiritual zero-point energy. So, as a connector or bridge between two worlds, is it possible to explain if the quantum communion can do something for unity and harmony of those two worlds?

Rein: Thank you. You want to go first, Bill, this time?

Keepin: Maybe I will go first, since I also want to address the second question of the previous speaker. In his second question, Dr. Richard Lewis inquired about dark energy. Taken together, dark matter and dark energy comprise something like 94 to 95% of the universe, which is therefore basically obscure to us, yet contains these deep mysteries. If we follow the epistemology of the heart for a moment, as understand it from spiritual masters and my own limited experience: as we plumb the depths of the heart, one veil after another is lifted. As each veil is lifted, a hidden truth is revealed. Then that truth is eventually seen to be another veil, and it is lifted, and yet another truth is revealed behind it, and so on.

So, I regard the dark energy and dark matter of the universe as areas of our ignorance—ignorance in a good sense, of the mysteries remaining to be explored about the nature of the universe. To me, it is very inspiring that so much of it is not understood at this point. This points to what David Bohm and Albert Einstein used to call "hidden variables"—the idea that there may be hidden processes, realms, and levels of reality that we do not understand, and cannot detect with our current scientific instrumentation, yet they are actually central to what is going on in the universe.

I was very struck by the presentation by Serge Kernbach, as well as the one by Arnaud Delorme, about the fact that sunspots are deeply connected with human experience. Also, the fact that living organisms seem to have this inexplicable capacity to pick out very weak signals from a whole field of environmental signals, and we have strong responses to these weak signals—and this is not well understood. It seems there is some important insight there about the nature of life itself, choosing what it is going to respond to, out of the field of variables that present themselves. In short, I regard dark energy and dark matter as a kind of reality check about how little we know, and a kind of meta-level of vast mysteries yet to be revealed.

Regarding the second question, my short answer is basically yes, "quantum communion" is the fundamental nature of the *connection* between matter, energy, and consciousness, and it unifies physical and 'spiritual' levels. I think that the pathway to discovering a First Cause will come in part from exploring the domains of unknowns through new kinds of epistemologies, which is one reason I am calling for the epistemology of the heart. So that is a brief, partial response. Thank you.

Rein: My answer to the question about the role of quantum fields in the brain mediating consciousness is, "Absolutely, Yes!" As I mentioned, there are 30-odd quantum theories out there, some of which involve quantum fields. My understanding, in essence, is that the quantum processes that occur in the brain—and more and more of them are beginning to be discovered, whether they are in microtubules, or in DNA, or any other biochemical process—they generate quantum fields. Quantum physicists tend to not discuss the role of quantum fields because they are obsessed with the particles rather than the fields, so the quantum fields tend to be left behind. I do agree that quantum fields are fundamental, and that they underlie and couple to ordinary electromagnetic fields. In fact, the relationship between quantum fields and electromagnetic fields is almost like a whole lecture in itself, and Menas could talk even better about that than I can. The idea is in terms of what is called "quantum control," which ultimately is an important concept to say that these fundamental quantum fields underlie and control the electromagnetic fields, which in turn underlie and control the biochemical processes in the brain.

Kafatos: Wonderful, thank you. Dr. Rein just made the connection to Dr. Keepin's point as well—extraordinary science. The next question is from Hayato Kondo. Please address the question to one of the speakers or commentators.

Hayato Kondo: Thank you very much for a wonderful session. I am a scientist in robotics engineering. So, it was a very eye-opening session. My question is for Dr. Rein. I liked your statement that the power of God is not far away, but it is everywhere. In that sense, my question is, besides the brain, is there any other system or mechanism to sense or resonate with the power of God? Do you have any thoughts on that?

Rein: That is kind of what I was trying to address. But if you are asking if there is a direct link between God and the brain, or if there are any direct resonances, I would say, "No!" I believe that God is such a subtle energy that His energy needs to be stepped down into the electromagnetic domain in order for the body or the brain to resonate with it, because it is at such a high, refined level. Therefore, I propose phase conjugation as a step-down mechanism whereby higher forms of energy can gradually be stepped down into the electromagnetic domain.

Kafatos: Thank you. Didier Guignard, you may now ask your question.

Didier Guignard: It is maybe for both main presenters because it is connected to things that they said. I was always fascinated by and interested in the Mandelbrot set and fractals, and now, upon hearing the presentations, I can make the connection between the holographic image and the fractal set. It means you are using something from a higher plane and bringing it down to a smaller dimension—smaller in size and in number of dimensions—and connecting that with quantum physics. Normally, when you have fractals, mathematically, they go down infinitely, but we know that reality does not go down infinitely. With quantum mechanics, you can have subtle variations that, at the individual level, can make something new and different. I could connect that with what Dr. Keepin said about different religions, about how they come to the same original point in a subtle, different way. So, after this comment, my question is whether, you—or maybe Dr. Rein—are aware of the work of Bruce Lipton, explaining that on the cellular level, there are basically the same functions as the whole organism, and that the membrane of the cell is connected with the function of the brain.

On the other hand, and I would like to have your comment on this issue, quantum physics allows our free will to exist. It means that through wave function collapse, for example—or we have to find some way other than wave function collapse—it is possible that there are connections that may or may not be made in the neurons, and in this way, our consciousness can work on our physical brain. One more step is that, through acting with the quantum fields and wave collapse, connections could even be directly from the level of God not only to our brain but to material objects as well, and could allow God to work in His creation without violating any law of His own creation.

Rein: Your turn, Bill. You can start.

Keepin: I will say a couple of things. As I understand it mathematically, the fractal is a nested set of self-similar structures. You are correct, of course, that there is a limit when we speak of fractals in the physical world—there is a limit of resolution beyond which physical fractal structures cannot go, but in the case of mathematics, the embedded miniature fractal replicas just continue scaling down, infinitesimally forever. Applying fractals in the consciousness domain, there may exist more generalized nested sets of similar structures in consciousness, and this enables us to understand how the finite human being can contain the infinity of consciousness. It is done in the same way that the single, tiny Mandelbrot replica contains within itself all the vast richness and intricate structure of the original, yet is hundreds of millions of times smaller. Similarly, the vast expanse and depths of cosmic consciousness is replicated in the mystical consciousness of a single human being, who on a physical level is billions of times smaller than the cosmos.

This principle is not just a property of fractals but also of real numbers. Mathematically, the finite can "contain" the infinite. For example, the cardinality (number of points) of a short line segment is equal to the cardinality of an N-dimensional space (for an arbitrarily large integer N). Each point in the N-space corresponds to a unique point in the segment. The entire infinite N-space can be fully "landed" onto the mathematical "airstrip" of a short line segment. My point is that this points to something analogous and fundamental in the very nature of consciousness.

Another point is about the origin of the Mandelbrot set: if you search for it inside the plot of the Mandelbrot set, you would have a hard time ever finding that equation. Instead, you find all these amazing intricate complexities and all this marvelous topology. The fact that all this derives from an extremely simple source is not at all evident in the plot and plane of the Mandelbrot set. There is no clear 'footprint' or image of the *source* of the fractal in the fractal itself. That source is on another ontological level, as it were. Similarly, I am suggesting that the Cause of the universe is not immediately evident in the manifest universe; the Cause is ontologically something very different, which we cannot perceive so clearly.

My sense is that there is a deep correlation between, let us say, the teachings in the scriptures of the Bhagavad Gita, that refers to the *bijam*, for example, and the Gospel of John that refers to the *logos*. In both cases, they say that the entire universe came out of this single seed principle. So, I am suggesting that there is a striking correlation between the way a simple formula gives rise to a vast manifestation of fractals in mathematics, and a simple principle of Divine consciousness gives rise to all of existence in both Eastern and Western spiritual traditions—and that is perhaps a deeply meaningful correspondence.

Rein: The speaker also brilliantly draws a parallel between the two lectures by drawing a parallel between the holographic dimension and the fractal dimension, in the sense that both are saying that every part of the whole contains the totality of the whole. So, in that sense, the body as a whole is composed of cells, and every cell, and every molecule in the cell, or every atom within every molecule contains the totality of the whole. That is, in essence, what we are talking about. If we view God as the whole, and humans as the smaller, individual components, then God is present within every aspect of the molecule, and the molecule is present within every aspect of the whole. It is a great parallel that I had not thought of, and I thank you very much for bringing it to our attention.

Keepin: If I may, I would just add one little line to that. If you think of a cell, it has its life through the heartbeat that pulses through the cell, and in this shared pulse, the cell is directly connected to the physical heart, which is vastly larger. This could have a spiritual parallel: the spiritual heart of each human being is directly connected to the spiritual pulse of the very heart of the Source of everything, or of God. So, there is a close parallel between the spiritual connection of the human soul to the heart of God, and the biological cell to the physical heart.

Rein: I could also add to what you had said, Bill, that the fundamental energy field that I have been talking about mostly in my talk would be love. It would be the fundamental energy at the quantum level, which links everything altogether.

Kafatos: Very good; excellent! This will be the last question, from Roumiana Tsenkova.

Roumiana Tsenkova: I have a question for Glen, and a comment on the discussion about fractality. We have proven fractality in water. When analyzing the water spectral data acquired under temperature changes, I can see the system of water molecules that make different structures. Those structures and water itself could be the denominator and the bridge between consciousness and fractality and holography. This is what I am writing about now—the research we are doing in this direction. It was very interesting to hear what Glen said about phase conjugation. I would like to ask if he could comment a little more on what is the connection between phase conjugation and where, exactly, is it in microtubules? How does it occur there?

Rein: I want to thank Roumiana for linking the theme of last year's conference with the theme of this year's conference, because last year, water was a big focus, at least what I talked about, and the idea that water has its own fractality to it. The fact is, of course, that God is present in everything, so God is present in water. I thank you for bringing both conferences together. The specific question about the relationship between phase conjugation and microtubules was not discussed in my talk because there is really no scientific connection. Theoretically speaking, yes, microtubules are a likely physicality of the phase conjugation mirror. Unfortunately, however, there is no science to support that phase conjugation occurs in microtubules. I brushed over that whole section because microtubules are so commonly discussed and there is a plethora of research on microtubules. Everyone in the consciousness field talks about microtubules, and I wanted to introduce something new. However, the connection between the two has yet to be determined. Hopefully, this conversation may inspire scientists who can measure phase conjugation in the lab to look at whether microtubules exhibit phase conjugation behavior. Microtubules do exhibit other types of quantum behavior, so it is a reasonable hypothesis, but as of yet no one has done the research.

Kafatos: Thank you all. I think this concludes our question-and-answer period. I would like to thank the speakers and commentators. In fact, this session was an asset to everyone participating because some really brilliant questions came up, which made some connections that were not thought of before. I would like to turn it over to the MC about the final session, but I want to give maybe a one-minute opportunity to either Dr. Keepin or Dr. Rein, if they wish to add something more—maybe some connections or anything else that came to their minds in this session. I want to give you the opportunity to say something, if you would like to do so.

Keepin: I will say just one thing about "zero energy"—the whole missing part about dark energy and dark matter. It may be deeply related to what in the mystical traditions is called the "dazzling darkness," and the notion that there is nothing but nothingness. So, there is this vast hidden depth at both the root and core of the mystical traditions and, it seems, at the root of quantum physics. To me, that is a very exciting frontier, where the two can meet, and these rich domains of the unknown are where the profound discoveries and new evolution will come from, in both science and spiritual consciousness.

Kafatos: Thank you so much.

Rein: Actually, I could add to that. If we are going to postulate that God is present in everything and everywhere, and that God is present in dark energy as well as in light energy, then, as scientists, it would behoove us to understand the relationship between dark energy and light energy, to help us ultimately understand how God manifests in the universe.

Kafatos: Yes. I would again like to thank everybody for your participation. ◊

ICSG II

CLOSING SESSION

Environmental Restoration: Hope from the Frontiers of Science

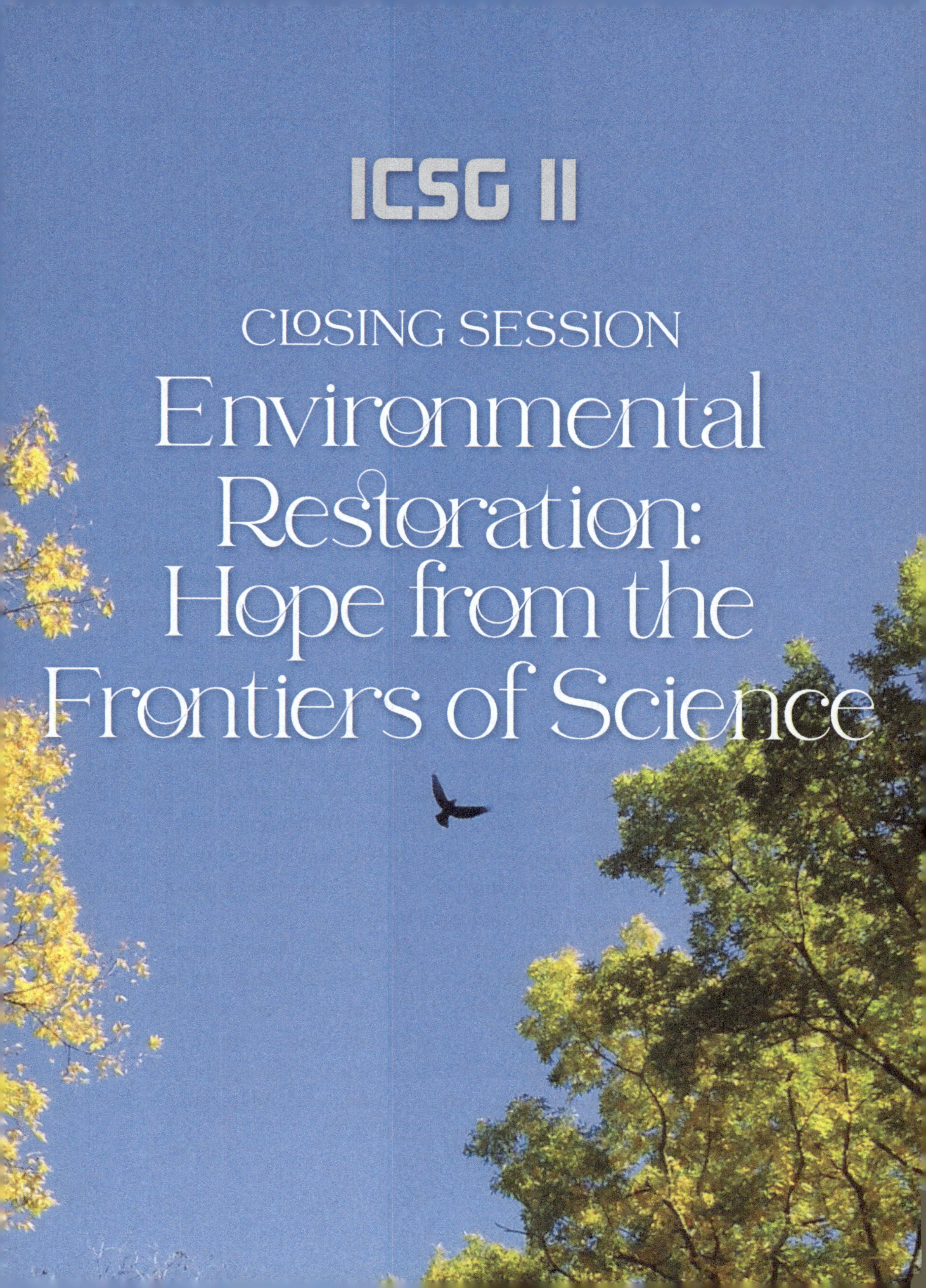

Closing Session | Report on Session 1

Unraveling the Secrets of the Universe: The Progress of Scientific History

Dr. Cassandra Vieten
Session 1 Chair

It was my great honor to chair Session 1 of this conference, which featured presentations by Prof. William Bengston and Dr. Dean Radin. Commentaries came from Dr. Margaret Moga, Dr. GunWoong Bahng, Dr. Helané Wahbeh, and Dr. Marjorie Woollacott. I would like to tell you a little bit about what happened during our session.

Dr. Bengston, first of all, is a Professor of Statistics and Research Methods at St. Joseph's College, New York, in the United States. He has been conducting research into anomalous healing for over 35 years and has done multiple animal experiments, across many universities, looking at a particular healing method that he co-developed and how it affects cancer and tumor shrinkage in mice. One interesting part of Dr. Bengston's presentation was that, through his research, he concluded that one of the main drivers of whether or not healing occurs from a healer to a mouse that had been given cancer artificially is that, first, biological need is required.

In other words, in his studies, when a healer directs information toward mice that do not have anything that needs to be healed, because they have not been injected with cancer, they do not show changes, whereas when there is a need, changes appear. One of the conclusions coming out of his research is that *need for healing* is an important part of the equation. He sees healing as an autonomic response to a biological need, and he proposes that when many healers and theorists report that they want to feel a subjective sense of intention and belief and spiritual connection, it is actually optional. This is controversial, in a way—this idea that conscious awareness of a spiritual connection is not actually needed for what is sometimes called spiritual healing or healing presence. He believes that it is an autonomic response to a biological need and that awareness is not necessary.

Having said that, while a subjective sense of connection in his view is not necessary, a biological connection does appear to be involved. So, there was speculation during his talk and throughout the session that this biological connection, (which in his paper he has talked about as coherence and brain frequencies, for example), could also find roots in a quantum connection between brains.

As we know, quantum theory and applied quantum research initially concluded that entanglement could happen only in very tiny and very cold settings, but over the years we have seen that photosynthesis has quantum mechanics involved, and the probability is that there are some quantum elements in our brain, so they may be involved. The other interesting thing is that even though a subjective connection does not seem to be necessary, intention is required. So, when we look at Bengston's paper, he says that when one walks down the street, they do intend to do so, but this has nothing to do with belief or sustained attention. Indeed, attentive walking, he says, will diminish efficacy. So, there is an initial intention, but there is almost a need to let go of that intention for the healing to happen. He says that the healing that he has observed does not conform to anything like a direct psychokinetic effect, but more like the "M5 model" that was proposed by Robert Jahn and Brenda Dunne—that it is a non-directed outcome.

He mentioned the apparent "storage" of healing in water, in cell medium, and in cotton. They have even done audio recordings of healing sessions and played them back, showing some future healing effect. This has led him to conclude that healing is more informational in nature as opposed to being energetic—or, if it involves energy, the energy is a vehicle for information.

Dr. Margaret Moga, associate professor of anatomy, cell biology, and physiology at the Indiana University School of Medicine, commented on this. She has worked with Dr. Bengston in the past. One very interesting thing that she and Dr. Bengston noticed was that there do seem to be magnetic field oscillations in response to healing, but rather than being close to the side of the healer, they are close to the side of the healee—that is, the human or the mouse receiving the healing. This could be very important, because weak magnetic fields have been shown by conventional research to slow cancer growth and improve survival. So, there may actually be a physicalist or materialist explanation, or at least reflection, of these healings.

Dr. Moga proposes what she calls a dual-level magnetic induction/extra dimensions model of energy healing. In other words, there is a physicalist materialist magnetic induction, but there is also healing coming from extra dimensions, or higher dimensions, and some humans may be able to access the higher dimensions based on their neuroanatomy.

Dr. GunWoong Bahng, a leading professor in mechanical engineering at the State University of New York in Korea, also commented on Dr. Bengston's presentation. He said that we need to pay attention to not only the quantity of healing energy, or purported healing energy, but also to the quality of the energy—it is not just about quantity or intensity, it is also about quality. He used the traditional five elements in Chinese medicine—water, wood, and so on—to say that these are the ways people in the past have framed the quality of energy.

This presentation was followed by Dr. Dean Radin, who is chief scientist at the Institute of Noetic Sciences and an associated distinguished professor of integral and transpersonal psychology at the California Institute of Integral Studies. Dean says that what is happening is that we are moving toward a postmaterialist worldview, one that views consciousness as fundamental, and many of these findings in parapsychology and energy healing over the past decades are leading us to the idea that physical or material reality may not be the dominant force, and that instead, consciousness may be fundamental to the nature of reality. He says that the scientific, technological, and sociological reverberations of such an ideological shift could dwarf all previous advancements in human knowledge.

He mentioned that there are these two problems: qualia and quanta. Qualia is related to the fact that any experience happens at all, which is sometimes referred to as the hard problem. Quanta is related to the fact that quantum objects seem to be sensitive to being observed. He calls this the even harder problem. He calls them two "clouds," referring to the two clouds from an earlier story of Lord Kelvin, and I encourage you to go back and watch his presentation to learn more about them. They are at the leading edge of anomalies that we are noticing that seem to challenge the current paradigm. This includes the phenomena of geniuses, savants, near-death experiences, mediumships, cases of reincarnation, and laboratory studies of psychic phenomena. The topic is seen as obviously controversial, and there is a lot of dismissing of the data, no matter how strong it becomes.

Dr. Radin proposes that the alternative to materialism might be called idealism. This is where consciousness is fundamental. From an idealistic perspective, these anomalies that we just mentioned are far easier to accommodate. They are not flying in the face of an idealistic paradigm, because in idealism, consciousness is not constrained by physical concepts like space, time, matter, or energy. If consciousness is not limited by these physical laws, then it is plausible that it is also not limited to gaining information only through the physical senses, nor is it limited to just the operations of the brain.

Dr. Marjorie Woollacott, a neuroscience professor at the University of Oregon, was a commentator. She mentioned that many experiences, such as near-death experiences and other mystical experiences that include nonlocal components—like out-of-body experiences or senses of unity—are associated with "ego dissolution," with an experience of the border between oneself and the external world dissolving. She proposes that possibly there is a reduction in activity in the default mode network that quiets the narrative mind. This results in reduction of the normal filtering process of the brain, thus potentially enabling an expanded awareness of subtler perceptions that would normally be filtered out and might give people access to information. Then, if information is across space and time from the future to the past or across great distances, if it is in fact, like Will Bengston said, the key to healing, as opposed to energy, that would really make sense—it could be the beginnings of a theory.

Finally, Dr. Helané Wahbeh, director of research at the Institute of Noetic Sciences, said that perhaps the difficulty is that the paradigm shift to postmaterialism does not affect just one scientific domain but all of them. So, there is a lot of resistance to such a broad, sweeping paradigm shift. Having consciousness as fundamental shifts our understanding of reality, and the implications would permeate every aspect of our lives, and therefore, it can become quite threatening. So, she said that if we hold the hypothetical assumption that consciousness is fundamental, we might want to focus on what we can learn about the mechanisms, mediators, moderators, and practical applications, as opposed to continuing to beat our heads against the wall to prove that it actually exists. It is better to look into how it works.

This was all followed by a lively discussion with questions from the audience, in the domains of energy healing and quantum reality. How can we explain any of this? I would say that several of the answers from the panelists were admittedly unsatisfactory, in the sense that even these people who have been studying these things for decades still do not really know where healing energy comes from. Is a quantum connection, when we talk about it in this way, more of a metaphor, or is it a plausible theory for how these things might be working?

So, I invite you to make sure that you view the videos and read the papers. It was a pleasure and an honor to chair this session, and I want to thank the conference organizers. ◊

Closing Session | Report on Session 2

The Promise of New Paradigm Science in Solving Environmental Problems

Professor Harald Walach
Session 2 Chair

Ladies and gentlemen:

We have had a very interesting session, with two quite different talks. The first, by Dr. Arnaud Delorme, was very challenging for traditional science because it pointed to the boundaries of what we can understand with the current naturalistic or materialist paradigm. It clearly pointed the way to seeing, maybe even demanding, that there should be a widening of the whole area of what the scope of science is because, as it is currently conceived, consciousness is produced by the brain, by the activity of the brain. That is the mainstream consensus in neuroscience and in many areas of biological and even medical science.

Now if you look at the data that Dr. Delorme presented, then our consciousness can obviously pick up information nonlocally, without causal signals. He presented work from mediums that were able to pick up information about people from the very images of them, whether they are dead or alive, which is very challenging. That leaves us with a more profound understanding of what consciousness really is. It reminds me of various developments within the science of consciousness over the last several years, which I have observed, that point to the fact that consciousness may not be produced by the brain but may be fundamental for our understanding of the world, and for our theory of matter, or the material world. Those empirical data are really important.

Prof. Jonathan Schooler, in his commentary, pointed out that the physical reality as we describe it now is not capturing our phenomenological experience. It does not say anything about our experience of our inner world. It does not say anything about the current "now," because the physical structure does not contain that. It does not say anything about the flow of time or changes. So, we need to realize that the worldview that science currently provides us with is incomplete. It may describe some aspects of reality quite well, and others, it has blanked out.

It does not make much sense to speak of a "scientific worldview" because there is none. There is always one approach towards the material world, from science, and that is maybe half, or even only a fifth or a tenth of reality that we can observe. Then there is another aspect, and another approach, which needs to respect consciousness, which we do not understand. That was for me the gist of that first presentation.

Prof. Won H. Kim, who also commented on Dr. Delorme's presentation, pointed to something very interesting—namely, what he calls *imaginary reality.* If you look at the formalism of quantum mechanics, the mathematics uses *i,* the square root of minus one, an imaginary number. Introducing i made the calculus of quantum mechanics possible. This is very fascinating because it produces some non-concrete entity that is needed for calculation. It is used by Prof. Kim to try to understand what he calls a holistic or holographic understanding of the world, and that may be at the root of what we do not understand.

The second presentation, by Dr. Serge Kernbach, came from a completely different angle, but it may have ended up at a crossroads. He was talking about biohybridization—how biological systems and technical systems merge in recent research in prostheses, intelligence, implants, interfaces between the brain and biological systems and technical systems. All these things have been at the forefront of scientific developments. If you look deeply into that, it points toward a potential merging of technical systems and biological systems. Now, all these developments are predicated on a very classical understanding of causality and the processes.

At the end of his talk, Dr. Kernbach pointed out that there may be quantum processes underlying that biology, and if that is so, our classical understanding will not be sufficient. The question is whether, for instance, nonlocal correlations or entanglement, even in macro systems, might be necessary to understand the very basic machinery of our biology. If you look at a very simple effect, such as photosynthesis, we know now, and we have been aware of it for a couple of years, that this is a quantum biological effect. It is a quantum effect in the macroscopic sphere. Accordingly, there may be quite a few processes that we do not understand that might be based on such macroscopic entanglement processes. If that is so, then causal analysis will be not sufficient to understand that and to engineer it.

So, we have a kind of meeting of these two developments in the middle, where perhaps quantum processes on both the macroscopic and microscopic planes play a decisive role. That was also borne out by one of the commentators, Prof. Roumiana Tsenkova, who looked at what she calls *aquaphotomics*—an area where the interactions between electromagnetic radiation, photon exchange, and water are at the center. If you look at that, then you also find a lot of holistic processes and very interesting phenomena that may not be easily understandable by very simple processes.

For me, all this boils down to whether a materialistic analysis that looks at only the outer aspects of the world—how biology works, how technical systems work, how they can interface—whether such a materialist paradigm is enough to understand life and, more importantly, to understand consciousness. My guess is, it is not, but we will know, maybe in 50, 100, or 200 years. Maybe we will not know, but we will have to make a decision. Maybe there is something like "optional ontology," meaning it is up to us how we want to view the world and how we want to view ourselves.

The Israeli historian Yuval Noah Harari has written a very interesting book, *Homo Deus*—or "Man as God." At the very end of that book, he points to the fact that our developments, such as those that Dr. Kernbach illustrated in his talk, lead to the conclusion that we have to decide as a society, as a culture, as human beings, whether we want to see ourselves as biocomputers, nothing more than simple biological machinery, or whether our consciousness—our inner experience, the way we feel, the way we intuit—plays a decisive role.

I think Dr. Delorme's talk pointed to the fact that it probably does, just from the phenomena, and Dr. Kernbach's talk pointed to the possibility that it might. Now, we, as researchers, as human beings interested in that interface between religion and science, should be aware of the fact that there has always been this intuition, even within science, that looking at the world only from the outside and not also from the inside of consciousness is only half the knowledge we can produce.

Alvin Plantinga, an important philosopher of religion, has written an interesting book titled, *Where the Conflict Really Lies: Science, Religion, and Naturalism.* It points to the fact that the real difference and the real conflict is not between religion and science, because religion has always been open to the findings of science even from the very start—it was the creator of science in the Middle Ages—and continues now. The conflict lies between a new type of religion, called *naturalism,* the scientific worldview, and science itself, because both science and true religion are open—open to the findings, open to the world, open to whatever there is.

Only if we predicate the world on a materialist paradigm and dedicate ourselves to a naturalistic religion, only then do we have to exclude those phenomena, such as those that Dr. Delorme talked about, and only then do we have to exclude consciousness as its own reality.

What I take from these two presentations is that maybe it is time to understand that this worldview, call it "neo-religion" or whatever you wish, is not enough for us. So, we have had very interesting insights into both the options and the limitations of that naturalistic (or materialist) paradigm. At the end, it raises this question: Where do we take the values, the moral orientation, the ethical guidance from, if not from our consciousness and its insights? If we do not do that, then we run a danger as a race, as human beings, of abolishing ourselves and our world. Therefore, there seems to be no way around considering consciousness as the hub of both science and religion. That might be another topic, perhaps for another talk or another conference.

Thank you for your attention. ◊

Closing Session | Report on Session 3

Ramifications of the Postmaterialist Perspective

Professor Menas Kafatos
Session 3 Chair

First, I would like to congratulate and thank Rev. Dr. Hak Ja Han Moon, the founder of this conference, as well as Dr. Douglas Joo, the conference chair, and Prof. Luc Montagnier, the conference co-chair. Last but not least, I thank the staff for the support we have received.

I will summarize the third session, which had the theme, *Ramifications of the Postmaterialist Perspective.* It sounds like a general subject, but we got quite particular in details. Some connections were made, and some of them actually resulted from the questions and answers period.

There were two main presentations. The first one was by Dr. Glen Rein, who spoke on the topic, *How Does Human Consciousness Interact with the Brain?* It is a very big issue, and he just touched the surface. However, I think what he touched upon is very crucial. He pointed out that in the current paradigm, the mind wants to generalize even consciousness, which is related to awareness beyond the mind, and to assign everything to neuronal processes, is a limiting way to move forward. In his talk, Rein concentrated on looking at the mind as some sort of field, an energy field, and he juxtaposed classical electromagnetic fields with quantum fields. They are related in the body, but they are not quite the same. One is according to quantum physics, and the other is more from electromagnetic theory. They are very closely connected.

What Glen Rein was specifically developing, which I find very interesting, is the whole idea that this mind-body duality may be very natural, although you usually think of it as violating classical physics, and so on. Perhaps the neural correlates, or the connections that are made in the brain, are more with purpose or function to basically enable the processes that occur in the brain. Of course, it is not just thinking—we think of thinking, but our mental functions are way more than that functionality. He looked at some of the alternative points of view, such as from Sir John Eccles and Karl Pribram.

Rein went into some molecular processes, looking at RNA and DNA as possible links to some of the things that are happening in the human body that are connected with the mind, if we talk about the mind in its extended sense, beyond just the thinking or logical mind. He particularly concentrated on the so-called *quantum phase conjugation.*

To illustrate what we are talking about here, think of a beam of light that is reflected from a mirror, where the "mirror" has to do equivalently with brain cells. The reflected beam is not simply following the laws of reflection, but it goes directly back to the source. So, that could represent some sort of mirroring in the brain, or a visual representation of dealing with the interactive universe. Of course, we know that a lot of psychology, or a lot of interaction with the universe, comes through the sense of vision that we have in the brain. In any case, there are some sort of quantum-like processes that take place.

Particularly in reporting about networks, Rein showed some of his network pictures that were very interesting. So, perhaps we could say that in some ways the mind is holographic, which would be in agreement with Karl Pribram's model. If that were the case, then maybe we are missing a big part of the picture, but in interdisciplinary approaches, that is where even more connections are made. As it turned out, the second speaker, Dr. William Keepin, also added a lot of connections.

Rein's talk was followed by two commentators. One of them was Dr. J. Scott Jordan, who is Professor and Chair in the Department of Psychology at Illinois State University and Director of the Institute for Prospective Cognition. He approached the subject from the viewpoint of a psychologist and could see that the psychological connections must also be addressed. He talked about a big issue in consciousness studies, namely, the issue of *qualia.*

He gave a somewhat humorous example by saying that he likes ice cream. So, he said, what is it about liking ice cream that is part of this whole picture? This led to a number of questions that were brought up again. I think the main point here is about contextuality—the context of an experience is very important. Usually, we think of neuronal processes as neurons firing, and maybe firing in unison, and so on. Then, what is the experience, what is the contextuality? I like, maybe, chocolate ice cream, or I like some other kind of ice cream. Where does that fit in the brain?

Then we had another commentator, Dr. Hiroshi Ishii. He is an MD and is the director of Kawasaki Kokoro Hospital in Japan. He had some very interesting questions. He said, if the brain and neurons are supportive of consciousness, then what can we actually say about the overall issue of scientific realism. How are the things that were brought up related to the so-called "weird" properties of quantum phenomena? It is often thought that because something looks out of the ordinary, it must be a quantum phenomenon. However, as a quantum physicist, I can say that quantum is not weird at all.

The second talk was given by Dr. William Keepin, who had participated in last year's conference as well. It was titled, *Can Investigations at the Frontiers of Science Lead to the First Cause?* As I said earlier, at first it did not seem to be connected to Dr. Rein's talk, but it turned out to be very connected.

Keepin's talk was very interesting because he brought up the more humane aspects that science (other than medical science) seems not to consider—that somehow science is cold, and somehow it is all external. He took exception to that in a very nice but forceful way. He has the ability to present big issues in a nice way so as not to offend anybody. Basically, he said that as we explore the frontiers of science, we open the door to invisible dimensions, such as those that we have been talking about in this conference, including some examples by Glen Rein. Maybe what we have now is supportive of the post-materialist perspective, and it may lead us to understand a host of phenomena.

He turned the tables on materialist scientists by presenting them with a big challenge, as they deny the reality of a lot of these phenomena. Their denial is, in a way, historical. I do not want to go into history, but perhaps some things were denied because the predominant point of view was going against them. There were religious leaders who suffered because of that, and he brought up some examples, not about the suffering but the common thinking of a lot of these religious thinkers and so-called dissidents—for example, people who held philosophical systems where the individual self and cosmic Self are closely connected.

Keepin gave the example of the Mandelbrot set, which is a simple mathematical way to show structures. They are not infinite because the Mandelbrot set eventually stops, but there are millions of points of resolution where, if you keep going through the Mandelbrot set and go through the very core, you eventually recover the whole thing. Keepin showed that graphically. So, that connects the holographic model in the brain with the Mandelbrot set. I would say that the Mandelbrot set is so dominant: we see it in plants, living forms, in some parts of animals, so called primitive animals. It is very interesting to contemplate.

He ended up saying something that I think was very profound. He rephrased the quote from Carl Sagan, who said that extraordinary claims need extraordinary proof. He pointed out that it is not the case here. So, are mainstream scientists ignoring the volumes of evidence that various speakers and writers talk about? He ended by making a passionate call for an "epistemology of love." It is very interesting that the word love shows up because I personally think it is really love, rather than particles, that moves the whole universe.

By the way, Glen Rein pointed out there is a tendency in physics today to just follow the particle view of the cosmos, and not the wave, because it is so much more concrete—you have properties of particles, such as spin and so on. So, the great success of the Standard Theory may also be a problem because it ignores or does not pay much attention to the wave aspect, which we know from the Copenhagen Interpretation or from the evolution of the Copenhagen Interpretation that it is a very important part of complementary reality.

Keepin's talk was followed by a commentary by Dr. Gary Schwartz, who was also the keynote speaker on the first day. He basically reemphasized the points made by Dr. Keepin, and he brought up some examples from his own writings. So, Dr. Schwartz was asking, do we seem to ignore or deny infinite love? That sounds like a really important question.

The second commentary was by Dr. Takahiro Hiroi, who is a senior research associate in planetary science at Brown University. He spoke about why the Earth is perhaps the right place for consciousness to develop. He pointed out that the Earth has a Moon that has profound effects in terms of stabilizing the weather and so on, preventing rapidly changing conditions. Also, fortuitously, the distance of the Moon from the Earth has allowed us to test a prediction of the theory of General Relativity during solar eclipses. There is some sort of cosmic thing going on out there, which is hard to imagine. He also talked about meteorites found in Antarctica. It is a fairly recent phenomenon, because Antarctica was not a frozen part of the planet. All this is very interesting.

I will finish by paying homage also to the audience, because some of these connections were pointed out or reinforced by the audience. For example, one question for Dr. Rein was whether the quantum phenomena he mentioned regarding phase conjugation may be much more general, perhaps having to do with the skin. Both the speaker and the commentator said that there are connections between the skin and the brain, so it is not just neurons telling us how to move and all that.

The other point here is, if indeed there is a field of love, maybe there is a connection also to dark matter and dark energy. I myself was originally an astronomer and astrophysicist, with an interest in exploring that. Essentially, it comes down to the "emptiness" or "darkness" in the sense of the divine, or the inner part of our heart that cannot be seen, and that is what brings everything together.

With that final comment, I would like to thank the organizers and recognize everyone who participated. Thank you very much. ◊

Closing Session

Concluding Remarks

Professor Luc Montagnier
Co-Chair, ICSG II

Good morning from Paris to our friends around the world. During this complicated time of pandemic, humanity has been thrown off balance. I myself encountered a surprising phenomenon first discovered by a colleague mathematician in the evolution of COVID-19; the existence of long Fibonacci series in the nucleotide sequences of the new variants of this virus, which are now emerging in the world human population.

Sadly for us, these variants used the flexibility of the Genetic Code to mutate in forms escaping antibodies raised by vaccines, while evolving harmoniously among themselves to maximize their growth.

'Don't die by ignorance' is the only intelligent response, and keep the lesson for the future. We have to be humble in unraveling the codes governing our biological existence.

Unfortunately, medical science has taken a dangerous turn. Scientific knowledge and clinical expertise are distorted at the world level by domination of an association of media, politicians and academic members.

Economic and financial interests surpass the health of our world community. We must restore the respect of scientific facts and protect the lives of all humans. I am pleased we have been able to spend these days together on zoom. A little science distances us from God; great science brings us closer to God.

Thank you for your kind attention.

And we are all most grateful to the months of hard work that the organizers have put into this important Second International Conference on Science and God.

I wish to extend my warmest appreciation to Rev. Dr. Madame Moon and all the Moon family for their continuing guidance and support. ◊

Closing Session

Closing Remarks

Dr. Douglas D.M. Joo
Chair, ICSG II

Respected Participants:

As we come to the end of the Second International Conference on Science and God (ICSG II), I wish to express my gratitude to all of you who have participated in this meeting.

This has been an unusual conference, presenting a unique set of challenges. Not only did we need to research the issues and topics and find expert speakers, we had to organize the meeting around different time zones spanning the globe. However, I was pleased to hear your comments that we raised the right issues at the right time. Also, each session was very productive, and I am very grateful for the tremendous enthusiasm and investment by all of you—co-chair, session chairs, speakers, commentators, and general participants, as well as the iPeaceTV technical team.

Thanks to you, I am happy to announce that the conference was another landmark event in the history of science. We at HJIFUS recognize and appreciate all your efforts.

It was as if we were riding on a space shuttle, and now we are about to land. We have not yet reached our final goal, but we have moved a step closer. We have come to recognize that scientific research has its limits, and science alone is not enough to protect and preserve the environment. Rather, human consciousness or attitude toward nature should be changed, guided by the Creator's heart.

At ICSG I last year, we overviewed the history of science and recognized that we are entering a new era of "frontier science," which includes a paradigm shift from a materialist to a postmaterialist perspective. We also heard about a variety of useful products derived from frontier science that can be used to restore our natural environment. In this manner, we formally opened the door to frontier science.

Now, as we end ICSG II, we can see that the presentations and commentaries have reaffirmed and given further substance to the viewpoint that we are at the threshold of another scientific revolution. A major feature of this revolution is the concept that consciousness, and not matter, is primary in the universe. In other words, science will soon move from a materialist framework to a post-materialist philosophical framework. You who are working at the frontiers of science are finding mountains of evidence in support of this revolution.

The study of consciousness has several ramifications. For example, it shows that the mind is not a simple product of the brain, but it interacts with the brain. Moreover, it shows that we are all interconnected with one another and with our natural environment. Once we realize this, we can have a genuine connection of heart with our surroundings, motivating us to exert proper stewardship over the environment and to live in peace with one another. This view reflects the perspective of the founders of this conference series.

Finally, it is clear that our experiences of the material world are leading us to understand nonmaterial reality, and eventually to recognize the First Cause of the universe, or God. Thus, we are at the unique, final stage in the history of science, and beginning a new history for humankind.

You are all heroes for contributing to a much brighter future for the world, and we have great hopes in your leadership. Let us develop our activities and find new projects that fit this new frontier science era. Also, please feel free to suggest topics we can cover during our next conference, ICSG III.

I am sorry that our time together was short, but I do hope to see you again next year, around this time.

Thank you very much. ◊

ICSG II

Appendix

Organizing Committee

Dr. Douglas D.M. Joo, Chair
Chair, HJIFUS

Dr. Dinshaw Dadachanji
Research Director, HJIFUS, USA

Mr. Didier Guignard
Founder and President,
Club Science Paris, France

Dr. Chul Hee Han
Professor, Chemistry,
Sun Moon University, Korea

Mr. Nobuyuki Iioka
Deputy Director, ICUS Committee, Japan

Dr. Sung Bae Jin
Chair, Hyo Jeong Academic Foundation, Korea

Dr. Frank Kaufmann
Director, Professors World Peace Academy, USA

Dr. Jin Choon Kim
Co-Chair, Cheon Il Guk Academy
of Arts and Sciences, Korea

Dr. Theodore Tadaaki Shimmyo
President Emeritus,
Unification Theological Seminary

Mr. Glenn Strait
Former Science Editor,
The World & I Magazine, USA

Dr. Frederick Swarts
Adjunct Associate Professor,
Sustainable Development,
University of Bridgeport, USA

Dr. Jong Choon Woo
Professor Emeritus, Department of Forestry,
Kangwon University, Korea

Master of Ceremonies

Dr. Tyler Hendricks
President Emeritus, Unification Theological Seminary, Korea

Staff

Mr. Sungwoog Hwang, Secretary-General, HJIFUS-Korea

Mr. Gregg Jones, Associate Research Director, HJIFUS, USA

Ms. Hi Seong Kim, Assistant Administrator, HJIFUS-Korea

Mr. Ryu-Sung Weinmann, Associate Administrative Director, HJIFUS, USA

* We wish to thank the many additional staff members who helped as volunteers for ICSG II.

Attendees List of ICSG II

Amemiya, Yoshiyuki
Professor Emeritus,
The University of Tokyo, Japan
PhD, Applied Physics and Engineering

Aoi, Kotaro
Graduate Student,
Kwansei Gakuin University, Japan
MA, Philosophy

Aoki, Kazuhiro
Professor,
Tokyo Medical and Dental University, Japan
PhD, Dentistry, Pharmacology

Bahng, GunWoong
Commentator, Session 1
Professor,
Department of Mechanical Engineering,
The State University of New York, Korea
PhD, Materials Science and Engineering

Bengston, William
Presenter, Session 1
President,
Society for Scientific Exploration, USA
PhD, Sociology

Cha, Jun Young
Former President,
The Segye Times, Korea
PhD, Literature

Choi, Jang-hoon
Pastor,
FFWPU Buyeo Church, Chungnam, Korea
PhD, Counseling Studies

Delorme, Arnaud
Presenter, Session 2
Associate Researcher,
Institute for Neural Computation,
University of California, San Diego, USA
PhD, Computational Neuroscience

Doroski, John
Former President and Executive Director,
The American Family Coalition, New York, USA

Famularo, Franco
President,
Universal Peace Federation, Canada
D.Min., Religion

Goldberg, Matthew
Pastor,
Maryland Family Church, USA

Guignard, Didier
Founder and President,
Club Science Paris, France
MS, Electronic and Computer Engineering

Hayashi, Masahisa
Professor Emeritus, Waseda University;
Representative Director,
Institute for Peace Policies, Japan
PhD, Public Finance

Hendricks, Tyler
Master of Ceremonies
President Emeritus,
Unification Theological Seminary, Korea
PhD, Religion

Hiroi, Takahiro
Commentator, Session 3
Senior Research Associate,
Brown University, USA
PhD, Mineralogy

Ishii, Hiroshi
Commentator, Session 3
Hospital Director,
Kawasaki Kokoro Hospital, Japan
MD, PhD, Geriatric Neuropsychology

Ishizaki, Junichi
Professor, Kobe Gakuin University, Japan
PhD, Clinical Psychology, Medicine

Jang, Seongdae
Pastor, FFWPU Boryung Church, Korea
PhD, Theology

Johnson, Mary
Former Foreign Service Officer, U.S. Department of State, USA

Joo, Douglas D.M.
Chair, ICSG II
Chair, Board of Directors; President, HJIFUS
Dr., Political Science

Jordan, J. Scott
Commentator, Session 3
Professor and Chair, Department of Psychology, Illinois State University, USA
PhD, Cognitive Psychology

Kafatos, Menas
Chair, Session 3
Fletcher Jones Endowed Chair Professor, Computational Physic, Chapman University, USA
PhD, Physics

Kato, Takashi
Professor, Graduate School, Chiba University, Japan
PhD, Theology, Biblical Studies

Keepin, William
Presenter, Session 3
Executive Director, Satyana Institute, Seattle USA and Vernon, Canada
PhD, Mathematical Physics

Kernbach, Serge
Presenter, Session 2
CEO, Cybres GmbH, Germany
Dr. rer. Nat., Electronic Engineering, Robotics

Kim, Jin Choon
Member, Board of Directors, HJIFUS-Korea
Professor, SunHak Universal Peace Graduate University, Korea
PhD, Physics

Kim, Won H.
Commentator, Session 2
Professor, Wonju College of Medicine, Yonsei University, Korea
PhD, Biochemistry

Kim, Dong Woo
Continental President, FFWPU South America, Brazil
PhD, Theology

Kim, Gongsoo
Former Pastor, FFWPU YoungKwang Church, Jeonnam, Korea
PhD, Pastoral Theology

Kondo, Hayato
Professor, Tokyo University of Marine Science and Technology, Japan
PhD, Engineering, Oceanography

Korotkov, Konstantin
Commentator, Session 2
President, International Union for Medical and Applied Bioelectrography, Israel
PhD, Quantum Physics, Biophysics

Kuribayashi, Minoru
Associate Professor, Graduate School, Okayama University, Japan
PhD, Communication Engineering

Lewis, Richard
Science Consultant, Unification Thought Institute, USA
PhD, Quantum Biochemistry

Lim, Jonghyuk
Senior Director, Office of Pastor and Ministry, FFWPU International Headquarters, Korea

McDevitt, Thomas
Chairman, *The Washington Times*, USA

Moffitt, Larry
Member, Board of Directors, HJIFUS-USA
Secretary General, Universal Peace Federation North America, USA;
MA, Communication

Moga, Margaret
Commentator, Session 1
Associate Professor,
Indiana University School of Medicine, USA
PhD, Neuroanatomy

Montagnier, Luc
Co-Chair, ICSG II
Founder and Director,
Le Fondation Luc Montagnier, Switzerland
MD, Medicine

Nagasawa, Kunihiro
Graduate Student,
The University of Tokyo, Japan

Ogasawara, Koji
Employee,
Toshiba Corporation, Japan
MA, Nuclear Engineering

Park, Guyong
Pastor,
FFWPU Boeun Church, Chungbuk, Korea
PhD, Theology

Park, Richard
Associate Research Scientist,
Yale University, USA
PhD, Biology

Phelps, Vicki
Member,
WFWP, USA

Radin, Dean
Presenter, Session 1
Chief Scientist,
Institute of Noetic Sciences, USA
PhD, Psychology

Rein, Glen
Presenter, Session 3
Founder and CEO,
Quantum-Biology Research, USA
PhD, Neurochemistry

Rigney, James
Founder, New Cosmic Spring LLC, USA
MA, Educational Administration

Rubik, Beverly
President and Founder,
Institute for Frontier Science, USA
PhD, Biophysics

Schooler, Jonathan
Commentator, Session 2
Professor, Psychological and Brain Sciences,
University of California Santa Barbara, USA
PhD, Psychology

Schwartz, Gary E.
Keynote, ICSG II; Commentator, Session 3
Professor, Psychology, Medicine,
Neurology, Psychiatry, and Surgery,
University of Arizona, USA
PhD, Psychology

Selover, Thomas
President,
SunHak UP Graduate University, Korea
Th.D., Comparative Religion

Shimmyo, Theodore
Member, Board of Directors, HJIFUS-USA
President Emeritus,
Unification Theological Seminary, USA
PhD, Theology

Shinoda, Yasunari
Graduate Student,
Tokyo Institute of Technology, Japan
MS, Hydrogen Energy

Strait, Glenn
Member, Board of Directors, HJIFUS-USA
Former Science Editor, *The World & I Magazine*

Swarts, Frederick
Member, Board of Directors, HJIFUS-USA
Adjunct Associate Professor, Sustainable Development, University of Bridgeport, USA
PhD, Biology

Takagi, Tetsuichi
Chief Researcher,
National Institute of Advanced Industrial Science and Technology, Japan
PhD, Geology

Tanimoto, Shoichi
Researcher,
Institute for Molecular Science, Japan
PhD, Theoretical Chemical

Togami, Yoshihiro
Professor,
Tezukayama Gakuin University, Japan
PhD, Electrical and Electronic Engineering

Tsenkova, Roumiana
Commentator, Session 2
Professor,
Kobe University, Japan
Dr. Eng., Dr. Agr.

Uhnak, Anton
Global Vice President,
Sun Moon University
East Europe, Czech Republic
MSc, Dr., Technical Engineering

Valone, Thomas
President,
Integrity Research Institute, USA
PhD, Engineering Physics

Vieten, Cassandra
Chair, Session 1
Director of Research,
Arthur C. Clarke Center for Human Imagination,
University of California, San Diego, USA
PhD, Clinical Psychology

Vitiello, Guiseppe
Honorary Professor,
Department of Physics "E.R. Caianiello,"
University of Salerno, Italy
PhD, Physics

Wahbeh, Helané
Commentator, Session 1
Director of Research,
Institute of Noetic Sciences, USA
ND, MCR, Naturopathic Medicine, Clinical Research

Walach, Harald
Chair, Session 2
Professor,
Poznan University of the Medical Sciences, Poland
PhD, Theory and History of Science, Clinical Psychology

Walton, Terry
Reverend, FFWPU, USA

Wilson, Andrew
Professor of Scripture Studies,
Unification Theological Seminary, USA
PhD, Near Eastern Languages and Civilizations

Woollacott, Marjorie
Commentator, Session 1
Professor,
Institute of Neuroscience,
University of Oregon, USA
PhD, Cell and Molecular Biology

Yamaguchi, Sachiko
Researcher, Institute for Peace Policies, Japan
PhD, Development Studies

Yamamoto, Shinnosuke
Employee, Boxyz, Inc., Japan

Yang, Susan
Professor of Computational Neuroscience,
Chapman University, USA
PhD, Computational Neuroscience

Founders of ICSG and HJIFUS

The Reverend Dr. Sun Myung Moon and the Reverend Dr. Hak Ja Han Moon began life in Holy Marriage in 1960. They based their life on the understanding that the family is the school of love and that all of humanity is one extended family, with God as our common Parent. They wholeheartedly dedicated themselves to practicing their teaching of true love and working to create a peaceful world. Faithful to this divine calling, they came to embody the philosophy of living for the sake of others and their mission as True Parents of humankind. They found that joyfully living for others resembles God's investment in the creation of the world and human beings. Living for the sake of others is the secret to resolving conflict and ending suffering in the world.

ICUS IV, 1975: Rev. Dr. Sun Myung Moon and Rev. Dr. Hak Ja Han Moon welcome participants in New York City.

Working side by side for more than 50 years, they founded and developed a vast array of organizations devoted to establishing peace, including the Holy Spirit Association for the Unification of World Christianity (HSA-UWC), Family Federation for World Peace and Unification

ICUS VIII, 1979: Rev. Dr. Moon delivers his Founder's Address at the Opening Plenary Session in Los Angeles.

(FFWPU), the Women's Federation for World Peace (WFWP), the Youth Federation for World Peace (YFWP), the Professors World Peace Academy (PWPA), and the Universal Peace Federation (UPF). They founded prominent media organizations such as *The Washington Times* and *The Segye Times,* among many others around the world. In addition, they established international institutions that support education, industry, sports, the arts, and humanitarian projects. Guided by principles that transcend differences between races, religions, and cultures, these organizations have been working to realize a world of interdependence, mutual prosperity, and universally shared values.

During the Cold War Era, the founders emphasized the need to resolve three core arenas of disorder in the world: communism, religious conflict, and the breakdown of the family. Based on their first-hand experiences in war-torn Korea, they saw that the militant atheism and class warfare advocated by communist forces had led to unspeakable atrocities and suffering of millions. In response, the founders built ideological and educational insti-

ICUS XIV, 1985: In front (*left to right*): Alvin Weinberg, Marcelo Alonso, Durwood Foster, Kenneth Mellanby, Sun Myung Moon, Hak Ja Han Moon, Morton A. Kaplan, Eugene Wigner, Chung Hwan Kwak, and Bo Hi Pak. In rear (*left to right*): W.W. Bartley, III, Tor Ragnar Gerholm, Huston Smith, Claude Villee, Alexander King, John Oxenham, Vicenzo Cappelletti, Jean Charon, and Neil A. Salonen.

tutions, such as the International Federation for Victory Over Communism (IFVOC), that challenged these forces and offered counterproposals that contributed significantly toward ending the Cold War.

They tackled head-on the dark history of religious conflicts through the International Religious Foundation (IRF), the American Clergy Leadership Conference (ACLC) and UPF's Middle East Peace Initiative (MEPI). By their example and sacrifice, they pioneered an ever-expanding federation of religious leaders working together in all parts of the world to foster a new era of religious harmony among believers from all faiths.

They taught that science is a necessary component in the pursuit of human happiness and progress. In 1972, they founded the International Conference on the Unity of the Sciences (ICUS). Based on their vision of interdisciplinary cooperation, ICUS conferences have brought together scien-

ICUS XXIII, 2017: Rev. Dr. Hak Ja Han Moon (*center*) speaks to ICUS XXIII participants about the responsibility of scientists to protect nature. Sitting with her are Dr. Sun Jin Moon (*left*), Chair of the Pacific Rim Education Foundation, and Dr. Yeon Ah Moon (*right*), Chair of Universal Peace Federation, Korea.

tists and scholars from the natural sciences, social sciences, and the humanities to work collaboratively with a holistic perspective pursuing the unity of the sciences with absolute values—that is, universal values and the fundamental cause and nature of the universe.

After the ascension of Sun Myung Moon in 2012, his wife, Hak Ja Han Moon, took responsibility for these projects with resolve. In honor of her husband's life, she established the Sunhak Peace Prize Foundation to recognize and award righteous heroes of our time who have taken the lead in the aspects of sustainable human development, conflict resolution, and ecological preservation. She introduced many new organizations such as the Hyo Jeong World Peace Foundation (HJWPF) and the World Christian Leadership Conference (WCLC) to expand on previous activities. *Hyo Jeong* means "filial heart"—indicating the teaching of a heart of filial piety toward God. She expanded UPF with initiatives like the International Association of Parliamentarians for Peace (IAPP), the International Summit Council for Peace (ISCP), the Interreligious Association for Peace and Development (IAPD), the International Association of Academicians for Peace (IAAP),

At the **Southeast Europe Peace Summit** organized by UPF, eight former heads of state from nations that had been in conflict during the Balkan wars receive ISCP founding medals from Rev. Dr. Hak Ja Han Moon (*center*), October 27, 2019, Tirana, Albania.

the International Media Association for Peace (IMAP), the International Association for Peace and Economic Development (IAED), and the Think Tank project which particularly gathers the wisdom and expertise of leaders in all sectors of society around the world who want to see the peaceful unification of the Korean Peninsula.

Among these new organizations, the Hyo Jeong International Foundation for the Unity of the Sciences (HJIFUS) was founded with the mission of solving environmental problems and living in harmony with nature. Rev. Dr. Hak Ja Han Moon has encouraged scientists and academics to pool their knowledge together to effectively address these issues. She has advocated the development of water purification systems for use in Africa and solving fine dust and air pollution in Eastern Asia. She expressed hope that the people of China, using modern technology, might transform the Gobi Desert into a verdant forest. The founders have expressed that to fulfill the mission of environmental protection and restoration, we need to clearly understand that the cosmos is like one body in which we are inseparably connected to the natural world and its fundamental cause. ◊

Introduction to HJIFUS

The Hyo Jeong International Foundation for the Unity of the Sciences (HJIFUS) is a nonprofit organization with the mission to protect and restore the Earth's natural environment. It was founded on the inspiration of Rev. Dr. Sun Myung Moon and Rev. Dr. Hak Ja Han Moon, who have elucidated a unique understanding of the relationships between God, humans, and nature.

Our current environmental problems include climate change; pollution of air, land, and water; changes in ecosystems; challenges in generating clean energy; and depletion of natural resources. These issues have detrimental consequences, such as deterioration in the quality of food and adverse effects on the health of humans and other species.

Modern science and technology are powerful tools that shape our world. When used competently and with the right intentions, they benefit both humans and nature. However, when misused through human selfishness, greed, or ignorance, they cause widespread environmental damage.

To help overcome environmental problems stemming from human behavior, HJIFUS promotes character education to improve the human attitude toward nature. On a practical level, the foundation identifies and helps implement well-rounded technological solutions and methods derived from both conventional science and frontier science. In turn, these technologies and how to use them are freely shared among all levels of society regardless of race, nationality, religion, social status, or culture. HJIFUS envisions a world where human beings live in transparency as one global family, based on the ideals of interdependence, mutual prosperity, and universally shared values. It is in this

ICUS XXIII participants with HJIFUS founder, Rev. Dr. Hak Ja Han Moon (*front row, center*), February 2017, Seoul, Korea.

way that environmental crises can ultimately be surmounted.

To accomplish these goals, HJIFUS works in six areas. One area of HJIFUS activity is the hosting of international environmental conferences. The International Conference on the Unity of the Sciences (ICUS) was originally started by the founders in 1972 to bring scientists and academics together pursuing the unity of the sciences with absolute values. The ICUS conferences were held regularly until the year 2000. From 2017, HJIFUS reconvened ICUS and has been hosting these conferences annually with a new focus to resolve environmental challenges. These meetings examine technologies derived from conventional science. In addition, HJIFUS has launched another conference series, beginning with the First International Conference on Science and God (ICSG I) in 2020. This series considers environmental solutions offered by frontier science, including postmaterialist science, which may represent a new dawn in the history of science.

In a second area, HJIFUS develops and offers environmental policy proposals to national and international agencies and policy makers. This is accomplished partly by distributing conference proceedings containing such

ICUS XXIV, February 2018, Seoul Korea: Rev. Dr. Hak Ja Han Moon (*right*) shares her vision with young scientists, encouraging them to lead future developments in protecting and restoring the Earth environment.

proposals. To broaden the scope of its activities and to collect consensus from experts and professionals around the world, HJIFUS is considering to establish a World Environmental Council or Congress.

In a third area, HJIFUS produces periodicals and other publications useful for educating the general public on the environment. On Earth Day, April 22, 2021, HJIFUS launched *The Earth & I,* a bimonthly online magazine that provides the latest news on local and global environmental issues, innovative technological solutions, advice on health and life, interviews with leading experts, and human-interest stories and testimonies related to improving the environment. Future publication plans include producing an encyclopedia of environmental science and technology, a *Who's Who* of key people and organizations in the environmental arena, a map of the global environment, and overviews of environmental policies in countries around the world.

The fourth area of HJIFUS activity involves leading social campaigns and programs to promote environmental restoration through service projects, hands-on education, and experiencing nature first-hand so that people

HJIFUS representatives visit **Senegal** and collect water samples for testing, June 2019.

of all ages can develop a caring and respectful relationship with the natural world. In July 2021, HJIFUS formed an outreach department to coordinate these activities and to explore partnerships with like-minded organizations.

The fifth area HJIFUS pursues is to raise the living conditions of people through practical projects. From 2018, HJIFUS has been laying the groundwork for a project that supplies devices that produce healthy drinking water for communities in need in Africa and elsewhere around the world.

Finally, HJIFUS supports the manufacture of everyday products that can substitute for items that exacerbate environmental problems. Through such products, people can live greener lifestyles and improve their health and well-being while taking care of the Earth.

With these approaches and multipronged strategies, HJIFUS is positioned to serve the international community with carefully thought-out plans, innovative technologies, and proven methods to protect, restore, and cherish the Earth for ages to come. HJIFUS organizations have been established in different parts of the world. They operate independently but share the same vision and mission. ◇

Boards of Directors

HJIFUS-Korea

(a nonprofit Korean organization)

Dr. Won Ju Chong McDevitt
Chief Secretary of Dr. Hak Ja Han Moon

Dr. Chul Hee Han
Professor, Chemistry, Sun Moon University

Dr. Sung Bae Jin
Chair,
Hyo Jeong Academic Foundation

Dr. Dong Moon Joo – *Chair*
President, HJIFUS-Korea

Dr. Jin Choon Kim
Co-Chair,
Cheon Il Guk Academy of Arts and Sciences

Dr. Yeon Ah Moon
Chair, Universal Peace Federation, Korea

Dr. Jong Choon Woo
Professor Emeritus,
Department of Forestry,
Kangwon National University

HJIFUS (USA)

(a U.S. 501(c)(3) organization)

Dr. Michael Jenkins
President,
Universal Peace Federation International

Dr. Douglas D.M. Joo – *Chair*
President, HJIFUS (USA)

Dr. Frank Kaufmann
Director, Professors World Peace Academy, USA

Mr. Larry Moffitt
Secretary General,
Universal Peace Federation, North America

Dr. Yeon Ah Moon
Chair, Universal Peace Federation, Korea

Dr. Theodore Tadaaki Shimmyo
President Emeritus,
Unification Theological Seminary

Mr. Glenn Strait
Former Editor, Natural Sciences,
The World & I Magazine

Dr. Frederick Allan Swarts
Adjunct Associate Professor,
Sustainable Development,
University of Bridgeport

Sponsoring and Supporting Organizations and Worldwide HJIFUS Offices

Sponsoring Organizations

Hyo Jeong International Foundation for the Unity of the Sciences–Korea
Hyo Jeong International Foundation for the Unity of the Sciences, USA

Supporting Organizations

The Washington Times Foundation
The Segye Times
Universal Peace Federation International
Family Federation for World Peace and Unification-International

Worldwide HJIFUS Offices

Korea
1st floor, 324-211 Misari-ro, Seorak-myeon Gapyeong-gun, Gyeonggi-do
Republic of Korea
Tel: +82-(0)70-7771-2604
E-mail: admin.kr@icus.org

United States
3600 New York Avenue, NE
Washington, DC 20002
USA
Tel: +1-202-636-2874
E-mail: admin@hjifus.org

Communication Offices

Japan
2nd Fl., Seiyaku Bldg.
5-13-2, Shinjuku, Shinjuku-ku
Tokyo 160-0022
Japan
Tel: +81-3-5362-0630
E-mail: world@pwpa-j.net

Europe
Espace Barrault
98 rue Barrault
75013 Paris
France
Tel: +33-6-68-31-63-87
E-mail: guignarddidier@gmail.com

Index

A

B

C

D

E

F

G

O

P

Q

R

S

T

UV

W

Made in United States
Orlando, FL
04 November 2026

90201495R00157